GEOGRAPHY

Heritage, Culture and Identity

Series Editor: Brian Graham,
School of Environmental Sciences, University of Ulster, UK

Other titles in this series

(Dis)Placing Empire
Renegotiating British Colonial Geographies
Edited by Lindsay J. Proudfoot and Michael M. Roche
ISBN 978 0 7546 4213 8

Preservation, Tourism and Nationalism
The Jewel of the German Past
Joshua Hagen
ISBN 978 0 7546 4324 1

Culture, Urbanism and Planning
Edited by Javier Monclus and Manuel Guardia
ISBN 978 0 7546 4623 5

Tradition, Culture and Development in Africa
Historical Lessons for Modern Development Planning
Ambe J. Njoh
ISBN 978 0 7546 4884 0

Heritage, Memory and the Politics of Identity
New Perspectives on the Cultural Landscape
Edited by Niamh Moore and Yvonne Whelan
ISBN 978 0 7546 4008 0

Geographies of Australian Heritages
Loving a Sunburnt Country?
Edited by Roy Jones and Brian J. Shaw
ISBN 978 0 7546 4858 1

Living Ruins, Value Conflicts
Argyro Loukaki
ISBN 978 0 7546 7228 X

Geography and Genealogy
Locating Personal Pasts

DALLEN J. TIMOTHY
Brigham Young University, Provo, Utah, USA

JEANNE KAY GUELKE
University of Waterloo, Ontario, Canada

LONDON AND NEW YORK

First published 2008 by Ashgate Publishing

2 Park Square, Milton Park, Abingdon, Oxon OX14 4RN
711 Third Avenue, New York, NY 10017, USA

Routledge is an imprint of the Taylor & Francis Group, an informa business

Copyright © Dallen J. Timothy and Jeanne Kay Guelke 2008

Dallen J. Timothy and Jeanne Kay Guelke have asserted their right under the Copyright, Designs and Patents Act, 1988, to be identified as the editors of this work.

All rights reserved. No part of this book may be reprinted or reproduced or utilised in any form or by any electronic, mechanical, or other means, now known or hereafter invented, including photocopying and recording, or in any information storage or retrieval system, without permission in writing from the publishers.

Notice:
Product or corporate names may be trademarks or registered trademarks, and are used only for identification and explanation without intent to infringe.

British Library Cataloguing in Publication Data
Geography and genealogy : locating personal pasts. -
   (Heritage, culture and identity)
   I. Timothy, Dallen J. II. Guelke, Jeanne Kay
   929.1

Library of Congress Cataloging-in-Publication Data
Geography and genealogy : locating personal pasts / edited by Dallen J. Timothy and Jeanne Kay Guelke.
   p. cm. -- (Heritage, culture, and identity)
   Includes bibliographical references and index.
   ISBN-13: 978-0-7546-7012-4
   ISBN-10: 0-7546-7012-0
   1. Genealogy. 2. Geography--Research--Methodology. 3. Historical geography --Methodology. 4. Geographic information systems. 5. Genealogy--Computer network resources. 6. Genealogy--Social aspects. 7. Information technology--Social aspects.
I. Timothy, Dallen J. II. Guelke, Jeanne Kay.

   CS14.G5 2007
   929'.1072--dc22

2007025296

ISBN 13: 978-0-7546-7012-4 (hbk)
ISBN 13: 978-1-138-26611-7 (pbk)

Contents

List of Figures and Tables		*vii*
1	Locating Personal Pasts: An Introduction *Jeanne Kay Guelke and Dallen J. Timothy*	1

PART I: TOOLS, SOURCES, AND IMPLICATIONS FOR GEOGRAPHY AND FAMILY HISTORY

2	The Unfolding Tale of Using Maps in Genealogical Research *Melinda Kashuba*	23
3	Genealogy, Historical Geography, and GIS: Parcel Mapping, Information Synergies, and Collaborative Opportunities *Mary B. Ruvane and G. Rebecca Dobbs*	43
4	A Genealogy of Environmental Impact Assessment *William Hunter*	63
5	Knitting the Transatlantic Bond: One Woman's Letters to America, 1860-1910 *Penny L. Richards*	83
6	Remaking Time and Space: The Internet, Digital Archives and Genealogy *Kevin Meethan*	99

PART II: GENEALOGY AS A CULTURAL PRACTICE

7	Genealogical Mobility: Tourism and the Search for a Personal Past *Dallen J. Timothy*	115
8	Genealogy as Religious Ritual: The Doctrine and Practice of Family History in the Church of Jesus Christ of Latter-day Saints *Samuel M. Otterstrom*	137
9	Genetics, Genealogy, and Geography *David C. Mountain and Jeanne Kay Guelke*	153
10	Conclusion: Personal Perspectives *Dallen J. Timothy and Jeanne Kay Guelke*	175
Index		*185*

List of Figures and Tables

Figures

Figure 2.1	Mean center of population in the United States, 1790-2000	24
Figure 2.2	Results from a place-name query of the GNIS database	33
Figure 3.1	Changes to "the old neighbourhood" in Louisville, Kentucky	44
Figure 3.2	GIS output in the form of spatial pattern analysis of land parcels	50
Figure 3.3	Large-scale GIS land parcel mapping based on archival sources	51
Figure 3.4	A typical 18th century survey from North Carolina's Granville District	53
Figure 3.5	18th century land records—links LAND to key relationships	55
Figure 3.6	Genealogical/historical material—links PEOPLE to key relationships	56
Figure 4.1	The Brothers Valley	71
Figure 4.2	Genealogy in place: The Walker Farmstead	72
Figure 4.3	The Swamp Creek Valley facing south	74
Figure 4.4	The Swamp Creek Valley historic district	76

Tables

Table 2.1	A survey of the number of lectures presented at U.S. genealogical conferences specifically related to maps and geographic sources	26
Table 2.2	The number of articles published in genealogical periodicals related to maps and geography between 1950-September 2006 (based upon PERSI version 8.2)	27

Chapter 1

Locating Personal Pasts: An Introduction

Jeanne Kay Guelke and Dallen J. Timothy

> Probably most people, most of the time, view the past not as a foreign but a deeply domestic realm. . . . History explores and explains pasts grown ever more opaque over time; heritage clarifies pasts so as to infuse them with present purposes. . . . Its many faults are inseparable from heritage's essential role in husbanding community, identity, continuity, indeed history itself. (David Lowenthal, *Possessed by the Past*, 1996: xi)

This book is about genealogy and its near-relation, family history, as cultural practices. If Lowenthal is correct, people's closest, most continuous domestic past is autobiographical. By extension, a personal heritage normally includes one's family and sense of continuity through ancestry, via one's birthright as a member of a kinship network, ethnic group, socio-economic class, and citizenry. Ancestral pasts thus live very much in the present for individuals. The exceptions almost prove the rule, because many adoptees today search for their natural parents; and people deprived of their sense of an ancestral history that has become "opaque over time" can become passionate about recovering it, as demonstrated in Alex Haley's (1976) bestseller, *Roots*. As a group of authors with social science backgrounds, predominantly in the discipline of geography, our twofold objective in this volume is to discuss how the geographer's research skills can inform the pursuit of family history, and also how understanding family history as a cultural practice can in turn generate new insights about society amongst geographers today.

This introductory chapter acquaints the reader with the outlines of family history as a leisure activity, and then situates family history within the context of contemporary human geography plus allied aspects of history and cognate social sciences. We argue that scholars in these disciplines who explore family history will find that family history in practice both illuminates and subverts key areas of interest, such as identity, landscapes of memory, and gate-keeping of knowledge. Because both the academy and popular family history should benefit from their mutual engagement, we conclude with recommendations for applied geographers, notably in the fields of geomatics, heritage resources, and tourism, to assist the conduct of family history through community outreach.

Genealogy for purposes of this paper is defined as the construction of family pedigrees: lists of ancestors and descendants. Family history includes genealogy, but also supplementary information about ancestors' lives and contexts. Both fit within Lowenthal's (1996: xi) above definition of heritage, but with the caveats that scholarly history, too, serves presentist purposes, and that the name "family history" is more firmly entrenched than "family heritage" at present. As a subset of heritage, family history relates closely to both the "domestic realm" and personal identity, and is also saturated with geographical themes such as residential location, immigration, diasporas, ethnicity, maps, heritage tourism, and homeland.

Family histories derive from a variety of sources: senior relatives' oral reminiscences; archived or on-line documents like emigration records; published histories of periods and places where ancestors lived or of significant events in which they participated; and family memorabilia like letters, personal effects, and photographs (see Richards, this volume). Some families add additional information or activities to their family histories, such as reunions involving extended families (Timothy, 1997), privately published chronicles of the family's generations, family web sites, and active networks of relatives and amateur genealogists researching and sharing their latest discoveries. Increasingly family historians turn to DNA testing through commercial "genetic genealogy" firms (Nash, 2006; see also Mountain and Guelke, this volume).

Genealogy of even a few decades ago seemed like the prerogative of traditional societies that emphasized kinship structures, such as in Mali and The Gambia of Haley's (1976) *Roots*; of European landed aristocrats whose pedigrees conferred jealously guarded privileges; or of descendants of "first families" of colonial societies, such as Pilgrims in the US. Family history today, in contrast, is fully integrated with western mass culture, and has become a popular leisure pastime (Rosenzweig and Thelen, 1998). "Ordinary" people happily search for their humble ancestors on the Internet, where once-hidden archival sources are now globally available on-line (Fryxell, 2003a). Special journals and finding aids are available for visible minorities, belying the now-outdated criticism that genealogy is just for whites or for the rich (cf. Carter-Smith and Croom, 2003; Ramirez, 2002).

Why be concerned with remote ancestors? Family history enthusiasts cannot be homogenized because their goals are not identical. Many engage in a quest for personal identity via their family past, in the belief that they express through their own lives some of their ancestral traits and values, or the results of their long-ago decisions, such as migration (Rosenzweig and Thelen, 1998: 45-52). Restated, in some sense they believe that they embody a family-based destiny. Family historian Russell Baker (1995: 28) exemplified the concept of personal identity defined by generational continuity: "I wanted my children to know that they were part of a long chain of humanity extending deep into the past and that they had some responsibility for extending it into the future." Others, alienated from contemporary urban society, may long for a sense of "blood and soil" ethnicity and of belonging to an ancestral homeland through claims of descent. Indeed, an ethnic group may be defined as "a collectivity of people who believe they share a common history, culture, or ancestry" (Scupin, 2003a: 67; Zelinski, 2001: 43-50).

Other rationales for family history activities appear to have little bearing on personal identity quests. Some amateur genealogists hope to locate distant but living relations in order to extend their family circles. Others enjoy the considerable puzzle-solving and "stamp collecting" aspects of finding one more ancestor through archival research. Professional genealogists search for other peoples' ancestors as a career (e.g. Mills, 2001, see also Kashuba, this volume).

Genealogy is often practiced as a subset of local history, and as such it attracts both amateur and professional historians (see Hunter, Chapter 4 and Ruvane and Dobbs, Chapter 3, this volume), many of whom participate in genealogical societies and journals. For example, the New England Historic Genealogical Society was founded in 1845 and publishes *The New England Historical and Genealogical Register*. Outspoken critics of fictionalized pedigrees in popular peerage compendia founded the first British genealogical societies and magazines, as a means of promoting factual accuracy (Camp, 1996).

The enormous collection of genealogical records of the Church of Jesus Christ of Latter-day Saints (Mormons) are available to the public, but principally serve Mormons' belief in proxy posthumous baptism and the reunification of their extended families after death (Allen et al., 1994/5; see also Otterstrom, this volume). Social and medical historians have long used genealogical records to examine kinship networks and transmission of hereditary diseases, respectively (Taylor and Crandall, 1986; Finkler, 2001). Many students do family history research simply because teachers require it for class projects designed to personalize broader historical events or social science principles (Maiewskij-Hay, 1999; White, 1998: 301; Friedenson and Rubin, 1984).

To skeptics, however, genealogy is an affectation of the gullible, who are bilked by unscrupulous commercial peddlers of spurious aristocratic surname histories, complete with coats of arms indiscriminately lifted from peerage compendia (Fowler, 2001; Camp, 1996; see Mills, 2001 on current professional genealogists' rigorous standards). Family history has been charged with being elitist due to its preoccupation with heraldry and "great men" in the family tree, homophobic due to its obsession with heterosexual norms, and sexist because it privileges the male line (Watson, 1989). Family history pursuits can be exclusionary, ethnocentric, and reactionary. White residents of the American South, for example, may express pride in having ancestors who enslaved African-Americans or served in the Confederate army and still believe in what they fought for (Shelton, 2001).

Critics of family history reason that an individual's self or identity cannot be a matter of blood quantum, of imaginative ties to a remote ancestral village that has dramatically evolved since one's forebears emigrated; or of locating progenitors in eighteenth century parish registers. Waters (1990, 2002) found ethnic identity for many assimilated white Americans to be a matter of childhood socialization or of personal active choices and individuation in the immediate present. This includes the choice of whether or not to identify with a particular ethnicity for millions of white Americans who are part of a homogenous mainstream (Waters, 1990). Obtaining a sense of one's roots through the family tree for such people is optional and unnecessary; but when pursued, it can lead people to unrealistic and romanticized beliefs about their personal heritage.

Scholarly studies of family history as a cultural practice in modern societies, perhaps as a consequence of its old reputation, are relatively scarce, notwithstanding the recent explosion of genealogy as a hobby and the emergence of genealogy-motivated tourism to archives and ancestral sites (McCain and Ray, 2003; Lambert,

2002; Bear, 2001; Louie, 2001; Grubgeld, 1997/98; Bruner, 1996; see also Timothy, this volume)[1]

Geographers' few forays into family history have been mostly negative. Environmental determinist Ellsworth Huntington used his own pedigree to support his now-discredited neo-Darwinian eugenics ideology (Martin, 1973: 188-90; Huntington, 1935). Lowenthal (1996: 196) depicted family history research as a merely amateurish heritage pursuit: an uninvited gate-crasher in the past, which is the rightful preserve of credentialed scholars. Nash (2004, 2005, 2006) critiqued genetic genealogy's proclivity to encourage misplaced ethnic identities, potentially to the detriment of sectarian conflicts of Britain and Ireland.

Although Catherine Nash's (2002) ground-breaking article "Genealogical Identities" critiqued naïve genealogical practice, she sympathetically described her interviews with genealogy tourists of Irish descent who visited Ireland in pursuit of archival records, a glimpse of an ancestral home, or meetings with newly-discovered relatives. Nash discussed with them issues of ancestry, ethnicity, nationality, identity, and a sense of belonging. While some of Nash's respondents seemed predictably unsophisticated or overly romantic about Ireland and their Irish roots, others unsentimentally discussed their hybrid ethnic backgrounds or discoveries of disgraced Irish ancestors. These interviewees confounded simplistic notions of a fixed, essentialized Irish identity, or of diasporic tourists retrieving a cohesive sense of self through traveling to an ancestral homeland.

Allied geographical research includes Gillian Rose's (2003) exploration of the meanings of family photographs to British women. This form of "visual culture" is an important artifact for family historians. Alison Blunt (2003) showed how genealogy was a crucial quest for Anglo-Indians, i.e. individuals with British paternal and Indian maternal lines, as they strove to obtain the privileges of British subjects by demonstrating British descent.

Timothy (1997) and Coles and Timothy's (2004) edited volume on "return" travel by members of diasporic communities present a more positive view of family history as a leisure choice. African Americans who travel to Africa typically find the experience to be intensely moving (Timothy and Teye, 2004) as do members of many other diasporic communities (see Basu, 2001; Morgan and Pritchard, 2004). On a more pecuniary level, British and Irish tourism bureaus now realize that "personal heritage tourism" or "roots tourism" can be lucrative for host communities, and actively promote it (see for example, www.goireland.com, and Morgan and Pritchard, 2004). As a geographer and professional genealogist, Melinda Kashuba (2005, see also Chapter 2, this volume) has written extensively on maps and geography as tools for family historians.

---

1   For examples of tourism companies specializing in custom tours for family reunions and for people who wish to visit ancestral locations, see any recent issue of *Family Tree Magazine* or other commercial family history magazine.

Theorizing Family History

The "old family history" could be justly criticized as a means of reifying essentialist concepts of fixed identity, socioeconomic class, homeland, and ethnicity (Peach, 2000) and as a discourse of insiders (e.g. British peers, descendants of early colonists) versus outsiders (the working poor, new immigrants). Most contemporary cultural geographers dislike ascribing material reality to concepts like class and ethnicity. These were too often used in the past as rationales for excluding people from a sense of belonging to particular places (Ashworth and Graham, 2005: 3) or to elite groups, or even for oppression of the poor and of visible minorities. The brand of genealogy that emerged in Britain during the Victorian era, for example, was concerned with reinforcing the pedigrees of the upper class, and demonstrating, however spuriously, that socially insecure middle class Britons willing to pay for their family trees also had aristocratic ancestors in the past (Camp, 1996). Genealogy became associated with reifying unequal class relations through the rubrics of noble blood, family seats, and coats-of-arms from medieval times—a heritage which most sensible people knew their ancestors never had.

Most Anglophone cultural geographers are also philosophically committed to the belief that cultural identities of various sorts are constantly negotiated through various power relations, whether these are imbedded institutions or individual practices (Duncan, 2006). Scholars of kinship today are more specifically concerned with *how* groups construct, represent, and police social boundaries in different situations and the power relations inherent in such practices (Franklin and McKinnon, 2000), rather than with rigid and naturalized categories like "family" and "ethnicity" or dismissive criticisms of these concepts.

Many social scientists believe that identities are fundamentally relational and situational, and that "individuals craft their identity through social performances, and hence that their identity is not a fixed essence" (Brodwin, 2002: 323). Victorian-style genealogy, based upon searches for aristocratic pedigrees as a means of asserting one's inherent superiority, was inadequate on these scores.

Yet essentialist or sentimental interpretations of ethnicity or identity via one's pedigree are by no means necessary to family history in practice. The "new family history" is typified by ordinary people who search census data posted on the Internet with no expectations of finding prestigious forebears. Family history today often reveals hybrid ethnicities and changing experiences of ethnicity over time, as well as collateral lore about lineages erased or exalted within family memories as one's relations responded to various conformist pressures in different locations and times (Domosh, 2002; Ohlinger, 1919; Tolzmann, 2000). Family history today is a reasonable fit with Battaglia's (1999: 115) concept of the "open identity" and her

> ... premise that persons, their subjectivities and identities (selves) are shaped by and shape relations to others, under the press of historical and cultural contingency ... selves are "not given to us" by natural law... not fixed or unchanging. And certainly they are not ontologically prior to relations of power ... selves are from the start an open question: subject to the constraints and manipulations of cultural forces ... [but are] capable, upon reflection, of breaking with and transforming the situations in which they are formed.

Just as one's citizenship and skin color at birth are involuntary and affect one's constraints or privileges within the contexts of place and period, family histories may reveal some of the contingencies and power relations that affect family narratives about their members' identities (Freeman, 2000; White, 1998; Haley, 1976).

Some of these power relations are institutional, such as Jim Crow laws in the American South that prohibited "miscegenation" and classified biracial children as African American no matter how few their African ancestors, or the minimum "blood quantum" requirements for tribal status among persons with some Native American descent (Elliott, 2003). Genealogy was crucial to the distinction between Aryan and Jew in Hitler's Germany. Another type of institution—marriage law—affects records kept for family historians' eventual retrieval, and also determines children's legal status and rights as legitimate or illegitimate.

Other power relations convey a sense of personal agency, such as great-grandparents' voluntary immigration decisions or the gate-keeping function of senior relatives who hold the keys to family memory (Richards, this volume), discussed below. Thus genealogy is not transparently revealed as a set of vital records or automatic expressions of social structures. One's relatives and ancestors may deliberately "prune their genealogies to further their concept of destiny, tidying themselves in order to infuse the past with the sense of rightness and purpose necessary to uphold the boundaries of the present" (Adams and Kasakoff, 1986: 57). Discoveries of such discourses operating within one's own family history narratives raise further questions about the kinds of structural pressures one's senior relations sought to address in thus selecting which stories to hand down to the future and which ones to withhold (Adams and Kasakoff, 1986: 57, 70), and the impact their decisions have had upon their progeny's identities in the present.

Place makes a big difference in family memory. Family history practice can appear very differently to "Old World" citizens, born and raised where they are surrounded by emblems of their ethnicity and relations on a daily basis; versus to structurally assimilated Americans, Canadians, Australians and other colonial descendents living where such emblems are either non-existent, highly imaginative, or are recoverable only after diligent searches for them (Waters, 1990).

Although critics would see a genealogy-based "identity quest" as misguided, the lack of interest in one's ancestry is equally problematic. The ahistorical concept of "the modern self-fashioned subject," who comes from nowhere, with individual talents and choices in place of ancestors, is, as Nash (2002: 36) aptly indicates, distinctly *modern*, even when that subject is understood as a relational self. As such, the ideology of the modern ahistorical self is a particular discourse that can be located in time and space (for example, as a liberal stance diffusing from the Enlightenment). One could also excavate the naturalized trope of evolutionary progress from a more tribal past to a more liberated individualistic present (Giddens, 1990, 1991). Thus reflexive explorations of "genealogical identities" may help to interrogate modern concepts of the self that today are developed within other agendas. For example, modern self-concepts based on identification with one's career, material possessions, personal appearance, or commercially-purveyed leisure activities serve hegemonic capitalist agendas, and are not self-evident improvements over other, and perhaps pre-modern, bases for grounding one's individuality (Giddens, 1991: 75, 197-200; Zelinski, 2001: 170-84), such as kinship or identification with a homeland.

Alienation in the Archives

Criticisms of genealogy as a leisure activity, moreover, may be based more on supposition than on actual practice. Hobbyists who pursue genealogy beyond a beginner level are generally aware of the pitfalls of presuming too much about their identities from genealogical evidence. Assuming, for example, that generations average about 25 years to produce a given offspring, then eight generations of ancestors will give an individual today 508 progenitors (assuming no endogamy and 100% retrieval of one's pedigree), not counting the various narratives that flesh out the genealogy skeleton. Analyzing one individual's family history becomes a massive and complex project that argues against presupposing much of a genetic inheritance from any one ancestor, even where "great people" emerge in one's pedigree. Most family historians know, or learn this mathematics.

Genealogical research itself can be frustrating to the point of alienating family historians from the very heritage they seek to discover. Despite the vast amount of on-line and print information available for family historians today, Anglophones with non-Anglophone ancestors may find their foreign language skills insufficient to the challenge of, say, reading adult-level German-language social history for places and periods of interest, as few of the sources have been translated into English. Even most native German-speakers feel unequal to the challenge of deciphering fuzzy microfilms of poorly preserved German parish records in obsolete handwritten scripts (Anderson and Thode, 2000). Genealogical research can also be expensive, if family historians pay professional genealogists or translators to do basic research beyond the level of their own archival or linguistic competence, or if they travel to overseas archives.

Identification with a particular family line or ethnicity is also complicated by today's political and cultural dissonances. For example, Americans and Anglo-Canadians with French Canadian ancestry may find a goldmine of information in the Francogene web site's extensive genealogical records for most of Québec's charter francophone families.[2] Québec was settled in the 17th and early 18th centuries by only about 350 families of French immigrants, who subsequently extensively inter-married, thus simplifying the task of locating ancestors via their surnames. During the 19th century, many Francophones immigrated to the northeastern United States in search of factory work, and some inter-married. Today many Anglo-Canadians bear French surnames, for similar reasons. Yet an Anglophone identity in the present may bar the English-speaking hybrid descendants of the Beauchamp or Beaudoin founding families from a warm welcome into the Québec *patrimonie*, due to longstanding tensions between the two language groups (Bear, 2001). France itself,

---

2 For a listing of Québec charter settlers on the Francogene website, see: <http://www.francogene.com/fichier.origine/index.php?l=en>.

back in the further mists of time, seemed even less plausible as a source of ancestral identification for "fusion" Anglophones.

Thus the notion that family history research inevitably plays into essentialist, nostalgic notions of personal heritage and belonging is unlikely to be the case in practice where language barriers work against tendencies to full-scale nostalgic ethnic identification in post-diasporic communities. They also suggest a range of power relations. The U.S. Constitution decrees no official language, yet the loss of non-English "heritage languages" among many native-born Americans speaks to the assimilation pressures experienced by their non-Anglophone ancestors.

Even without language difficulties, family historians often encounter frustrations of missing data, illegible microfilms, and cemeteries re-landscaped. Quests for specific ancestors may lead to an impenetrable "brick wall". Multiple surname spellings, pronunciations, typographical errors in transcriptions, and name-changes in the non-standardized practices of the past can make tracing bloodlines ambiguous (Anderson and Thode, 2000). Conducting genealogical research offers no guarantee that the records will be found upon which to base any kind of emotional tie other than one of frustration.

The "doing" of family history also reveals the range of power relations endemic to historical research of any kind. There are the standard archivist issues of who decides what information to record about whom; and what information to accession, preserve, and promote to the public (Meethan, this volume). Senior relatives become gate-keepers, choosing whether to let out skeletons from the family closet—as they define them (Richards, this volume).

Alienation can also appear when newly discovered sourcebooks on family surnames happen to translate one's ancestral surnames as meaning "toothless" or "quarrelsome person". Anyone of European descent who is politically incorrect enough to bask in the pedigree peddlers' write-ups of the Norman conquest, family seats, coats of arms, and the like, may be in for a rude shock when locating more immediate ancestors in census rolls that record them as illiterate day-laborers, or when social histories reveal that legitimately-claimed aristocrats in one's family tree ruthlessly exploited their peasants in order to sustain lavish lifestyles.

Genealogy is normally a tracing of ancestors by surname. Yet Scott, Tehranian, and Mathias (2002) point out that permanent surnames were by no means indigenous cultural practices in many parts of the world, including medieval Europe, and that permanent family names were often forced upon the proletariat by ruling classes or the state for such purposes as taxation or determining descendants' social degree, and hence statutory privileges or inequality. Patronyms that changed with each generation in Scandinavia, Wales, and areas of Friesian settlement in northern Europe, for example, were common into the 19th century. Thus surname histories may reveal to the geographer the power relations of how state and military conquest can extinguish minorities' rights to retain their cultural practices. Yet the discovery of patronymics in one's family tree might not satisfy either a desire for identification with a durable and venerable family surname nor for easy work in the archives.

Is genealogy sexist as critics have charged, given the structurally greater ease (and oftentimes the hobbyist's desire) of tracing paternal surnames through the generations? Our experience is that the state of preservation and accessibility of vital records is no respecter of gender; female lines may be easier to trace where better data are available for them than for males, or where female lines tended to stay in the same site while men migrated—and hence effectively opted out of a given data base. The majority of one's ancestors are mixed between male and female lines, in any event (one's mother's father's mother, and so on). For example, eight great-parents include only one individual apiece traceable through a strictly paternal or maternal line. To be sure, it is hard to get around the prevalent European custom of women taking their fathers' and husbands' surnames. Yet societies in which women keep their birth surnames (e.g. Korea) are not demonstrably pro-feminist.

Battaglia's (1999: 115) definition of the open identity cited above, with its insistence on mixtures of relationships, contingencies, power relations, and personal choice, thus seems appropriate to the study of family history. Her themes will be developed in the next section in terms of what family history can reveal about identities based upon religious affiliation, socioeconomic class, ethnicity, and place-based heritage.

Culture, Place, and the Problem of Family History

Arguing that the leisure pursuit of family history is just as likely to undermine, as it is to reinforce essentialist markers of identity, is not to argue that societies believe that social categories are pure fabrications and dismiss them accordingly. Ancestry is a crucial defining feature of personal and ethnic identity to people in many societies today, even in modern Europe (Iossifides, 1991: 137). It defines marriage fields and taboos, inheritance law, genetic predispositions to certain illnesses, and oftentimes socio-economic class for even the most "self-fashioned" of individuals. Ethnicity and clan membership even today may be reckoned by descent from a legendary common ancestor or founding charter group, from which members presumably trace their descent (Scupin, 2003a: 67). In this manner, for example, contemporary Jewish identities are traced back to Abraham and Sarah in the biblical book of Genesis, and Acadians of eastern Canada base their identity upon descent from a small charter group of immigrants as well as a sense of shared history and common mother tongue. In political hot-spots with long historical memories, such as Northern Ireland, Israel/Palestine, and the Balkan Peninsula, ancestors' heroic deeds or victimization serve, together with religion, to reinforce notions of group insiders and outsiders, of countrymen and enemies linked or separated through ties of historical experience that are fundamentally derived from one's ancestry.

Ancestry is a fundamental basis for citizenship in nations that wish to limit immigration or, alternatively, to repatriate individuals with the accepted ethnic identity. For example, many third-generation German residents of Turkish descent were ineligible for German citizenship until 2000 (Canefe, 1998), whereas claiming one Irish grandparent may qualify foreigners for Irish citizenship (Embassy of Ireland, 2001). Ancestry defines the parameters of contested space in volatile political hotspots: who has a right to occupy a particular place by virtue of birthright, and who is an interloper or "new immigrant." Ancestry reinforces minority categories in racist societies. Even critics of genealogy nevertheless experience, as citizens of a particular nation, the results of decisions made by their ancestors about whether to stay within their homeland or to emigrate. In our physical appearance human beings express the outcomes of our ancestors' genes, sexuality, and childbearing. Some individuals live daily with the results of involuntary events experienced by their ancestors, such as forced migrations or hereditary health predispositions.

By this logic, the self cannot be constrained to its embodied, presentist space-time paths, framed by dates of the individual's birth and death but is a dynamic and complex continuum that originates at some unknowable date in the past before one's birth, and that will continue dynamically through descendents or influences upon others after one's death. We say this not as an apologia for family history pursuits, nor in promotion of a super-organic definition of identity, but merely to illustrate family history's subversive potential for modern identities based narrowly upon personal achievement, conspicuous consumption, or current group affiliations.

In the matter of religion, membership in a particular faith runs very deep into the past for many adherents. The Roman Catholic Church, for example, has profound ties to what it means to be of Polish, Italian, or Mexican descent for many members of these nationalities, and may be a highly salient aspect of Catholics' personal identity. The majority of today's practicing Roman Catholics did not convert to Catholicism but observe it as a legacy from their ancestors. What it means to have a particular kinship network may be expressed in attendance at church-based confirmations, weddings, and funerals of relatives.

Similarly, what it means not to have religious links to ethnicity and kinship may equally run back several generations, with the rise of secularism in the 19th century. Even in a modern secular state, decision on whether or not to bring up a child in a particular faith—or in no faith at all—may run back several generations' worth of habituated attitudes, and cannot entirely be naturalized as "independent decisions" taken by self-fashioned individuals.

A truism of historical geography is that the "working poor" are the most likely group to emigrate overseas. Consequently, family historians living in former colonies are far more likely to find such individuals, and even illiterate ditch-diggers, in their pedigrees than aristocratic land owners. Colonial Australia and Georgia in the American South began as penal colonies (Lambert, 2002)—a former embarrassment to some descendants of these early "first families".

Recovery of ancestors can consequently facilitate more conscious interrogation of the construction and representations of one's own class origins, socio-economic mobility, and identities. Sometimes a neglected branch of a family's past lies buried in order to reinforce the family's desired self-image in the present. The easy dismissal of family history interests as wrong-headed might equally stem from modern commitments to upward mobility and forging a more prestigious socio-economic identity.

Ethnicities can become prioritized within families in similar ways. The social history of French Canadians who migrated to the 19th century mill towns of New England reveals the region's ethnic pecking order; Puritan-descended Yankee English felt themselves to be superior to the Irish-Americans, and the Irish in turn claimed superiority over French Canadians (Brault, 1986). This ordering of ethnic prejudices occurred in part because each wave of immigrants was successively blamed by its predecessor for under-cutting wages in the local mills. Social pressures for ethnic or class erasures among intermarried families could easily influence them to identify with the more socially prestigious side of their ancestry. Being on the wrong side of two world wars, for example, convinced many Americans of German descent to downplay their German ancestry, to the extent that today many Americans with German forebears have no idea where in Germany their ancestors originated (Tolzmann, 2000; Trommler and McVeigh, 1985).

Of course, one could learn about citizenship or social mobility simply by reading social science or history books about anyone. The difference, we think, in interrogating one's own family history, is that taken-for-granted assumptions about one's own identity are now at risk. Herein lies just part of the radical potential of family history.

Homeland and Nationality

For many family historians, their practice strengthens their sense of ties to a particular place. Although some family historians express little interest in the locations of their ancestors (either they live in the ancestral locations to begin with, or the places where their forebears lived simply do not resonate with them), Nash (2002) found that the desire to experience their ancestral homeland is a strong motivator for genealogy tourists who visit Ireland. Similarly, patriotic third-generation (or more) Canadians proudly wear their Scottish clan's tartans on Robert Burns's birthday, even though some social scientists now believe that many Scottish family tartans are a relatively late invention, originating as recently as the Victorian era (Cobley, 2004; Hobsbawm and Ranger, 1992) The prospect of discovering African roots in a particular African homeland has been a magnetic lure in African American family history and tourism to Africa (Timothy and Teye, 2004; Berry and Henderson, 2002; Bruner, 1996; Haley, 1972). Family historians are known to scour maps in order to locate the communities of remote ancestors, to plan their vacations around visits to their progenitors' grave sites, and to collect memorabilia from these locations (Timothy, 1997; see also Chapter 7, this volume). A propensity for genealogical research thus typically entails considerable geography, whether actualized or imaginative.

Ancestry can equally effect a sense of belonging to different regions of a genealogist's own country. No doubt most places have their "first families" and "old names" derived from settler progenitors. In the United States, having male ancestors who fought in the Civil War confers social status and a sense of belonging in the South, but not particularly in the northern or western states. Descendants of the Pilgrims who traveled to Plymouth, Massachusetts, on the *Mayflower* in 1620 live all over the world today, yet trace their antecedents to one pivotal time and place in U.S. history. Organizations like Daughters of the American Revolution were (and to a more limited extent remain) a means for socially-conscious Anglo residents of the eastern seaboard to mark particular historic times and places as defining the "real" Americans, in contrast to new immigrants and their offspring.

The search for an ancestral homeland through what Timothy (1997) terms personal heritage tourism can also have destabilizing effects. Rootlessness, or at least residential mobility, is typical of many family trees. Because vital records normally are found in particular communities, such as parish registers, they can reveal peripatetic tendencies among families. Family history can also reveal where places themselves are unstable. For example, any "dot on a map" that one could assign to Dutch ancestors' origins might be ephemeral where ancestors repeatedly settled on lands newly claimed from the sea. Sometimes forebears came from "somewhere else" as far back as family historians have been able to trace them.

Several points could be made from these examples about *Heimat* or *Blut und Boden* (blood and soil) attachment to an ancestral homeland or national identity.[3] The first question for individuals with multi-ethnic backgrounds is: with which homeland or nationality does one identify? And in the absence of the sorts of rules that govern some traditional societies, on what basis could such a selection be made? Second is that many family historians have never, or perhaps only briefly, visited as a foreign tourist any of the communities where their ancestors lived. Third, Old World communities function in today's modern world, and not in some kind of time warp retaining the century that one's ancestors inhabited. *Heimat* is a moving target. Fourth, if one were to claim more than a sentimental attachment to his or her landscapes, would the local residents return the favor? Or would they "police the boundaries" of legitimate ties to their district and exclude genealogy tourists as outsiders? This has been a challenge for African Americans who visit slave sites in West Africa. They go to Ghana and other countries expecting to be welcomed "home" by the Africans of today but often leave disappointed and hurt because they were treated and "processed" simply as foreign tourists (Hasty, 2002; Timothy and Teye, 2004). Fifth, few family historians know much about emigrant ancestors' own feelings of attachment or bitterness towards their natal soil, with the exception of a few tight-knit groups that encouraged members to record their memoirs, such as the Mormons. It is likely that ancestors of family historians said, "Good riddance" to the same village that their modern descendants now might wish to sentimentalize. Thus the "new genealogy" actively creates new place identities and social relations.

Textbook maps delineating ethnic homelands seem clear enough, but the diagnostic attributes of particular ethnic groups and rationales for drawing the boundaries around them might vary with each ethnic group. The Francogene website mentioned above clarifies that most French Canadians are actually related to one another owing to a small and previously isolated charter population. This is not the case in southwestern Ireland or Germany due to frequent migrations. Language and religion seem clear enough in establishing place-based identities and ethnicities, yet sometimes populations are pressured under disadvantageous power relations to adopt official languages (such as Low German speakers learning High German) or state religions (Iberian Jews who converted to Catholicism) in order to erase local ethnic attachments, not as an expression of them.

In any event, ethnic group boundaries may be quite permeable from the perspective of recent human genome projects. Research on genetic distances

---

3   The idea of *Heimat* or a Germanic homeland experienced at flexible scales of home, district, or nation, replete with emotions of sentimentality and lost innocence, is described in Blickle (2002).

reveals, for example, that the Netherlands and Denmark show a high degree of genetic similarity regardless of their linguistic, political, and spatial divergence (Cavalli-Sforza et al., 1994: 270). Defining nationality on the basis of the ancestry of the inhabitants serves contemporary social and political purposes, but is strongly supported by genealogical or genetic evidence only in cases of geographically or culturally isolated endogamous groups (Molnar, 1998: 286-7; see also Mountain and Guelke, this volume). Thus the interesting questions for cultural geographers become, as cited above, how and on what basis social groups nevertheless define themselves, police their boundaries, and establish the criteria for acceptance of hybrid individuals as group members (Bhabha, 1994; Brodwin, 2002; Franklin and McKinnon, 2000).

Prospects for Geography and Genealogy

We have tried to show in this introduction how genealogical research can subvert essentialist and static categories of personal identity that are basic to human geography today, such as socio-economic class, religion, ethnicity, nationality, and identification with a homeland. Kinship and the family are venerable analytical categories in anthropology and sociology, but are little explored in geography. We argue, further, that these themes are best explored when academics undertake their own family histories (see Freeman, 2000; White, 1998) or work as a matter of applied scholarship with people who do. Of course family history research is not a prerequisite for reaching such normative postmodern conclusions, but when people undertake family history research, these "lessons" are no longer depersonalized or abstract. Reflexive questions (Moss, 2001) about the researcher's own family history can move her or his identity from the safe and obscured margins of the research enterprise to its vulnerable—and hence transformative—center.

There are, however, dangers in pinning identities upon today's critical social history and historical geography. The danger in revisionist family history is in simply replacing one naturalized narrative and set of archetypes (historical heroes, villains, and victims) with others that are equally cloaked with pretenses and ulterior motives. If an Irish American cannot demonstrate descent from a particular county's "principal inhabitants" (Morris, 1992) under the Old Genealogy, for example, why not adopt valorizing tropes about salt-of-the-earth Irish tenants, the dispossessed rightful heirs to the land, whom the "best families" so wickedly oppressed? Or why not script those victims as valiantly resisting oppression while gratefully receiving their liberation at the hands of the present Irish state? Such revisionist history plots are equally open to contest and debate.

Ideally, the radical potential of family history runs deeper than replacing Narrative A with Narrative B (Lambert, 2000). A peripatetic family history could cause the researcher to question not only the basis for an ethnic identity and homeland, but indeed, whether he or she has an ethnicity in the conventional sense (Scupin, 2003b; Waters, 2002, 1990). In today's academic climate, ethnicity has become synonymous with "visible minority" or "new immigrant" and is increasingly reduced to skin color, language barriers, or immigration card status (Monk, 2002; Rumbaut and Fortes,

2001; McKee, 2000; Yang, 2000). Euro-Americans are increasingly represented merely as perpetrators of racism, segregation, and white supremacy movements against people of color, as foils for people who do have an ethnicity. Long-term family histories might reveal whether Anglophone white America has virtually achieved status as an ethnicity in recent years, using definitions based upon a sense of shared history or cultural coherence rather than ancestry (Berry and Henderson, 2002; Zelinsky, 2001: 28-38) or whether white Americans inhabit a hybrid or post-ethnic space.

Scholarly approaches to family history research can make a difference to this dilemma of Anglo American ethnicity. Postcolonial scholars have condemned the discourse of "whiteness as an unmarked category" in defining colonial relationships, a point somewhat blunted by research on ethnic and religious cleavages within groups of northern European descent. Family history has the potential to unpack further the concept of "whiteness as an unmarked category" by showing precisely how "whiteness" constructs itself in the hands of ancestors who suppressed unwanted aspects of their own hybridity before turning their gaze on colonized people of color.

Bernard Anderson (1991) talks about how the formation of the state requires people to agree to believe in "imagined communities". His concept applied to family history can reveal that different groups have historically imagined themselves on very different terms. The popularization of family history via the Internet further promotes new images of ancestral ties to place and ethnic identifications.

As genealogy tourism increases in popularity, it also suggests a rethinking of the heritage tourism industry and its geography, with a fissioning of destination sites as visitors seek ancestral dwellings, neighborhoods, and local libraries versus specially designated and marketed mass heritage sites (Timothy, 1997; see also Chapter 7, this volume). For example, genealogy tourists of Scottish descent may be less inclined to seek their Scottish identity in national monuments commemorating famous battles (Edensor, 2002) than through their great-grandparents' street addresses and the cemeteries where they lie buried (Basu, 2001). Genealogical researchers on a personal identity quest may show little interest in the mass marketing agendas of commercial tour operators or tourism ministries. The emergence of niche market service providers for genealogical tourists, however, appears in advertisements in family history magazines, raising interesting questions about how the tour operators work with such individualized itineraries that might take a single family of clients far from normal tourist destinations.

As a popular form of research conducted by "ordinary" people without proper scholarly credentials, the current widespread family history phenomenon should also destabilize academicians' non-reflexive restriction of "true" research to the elect, and their gate-keeping determinations of who is "in place" or "out of place" in archives and research libraries (Rosenzweig and Thelen, 1998; see also Richards, Chapter 5 and Meethan, Chapter 6, this volume).

Rosenzweig and Thelen (1998) conducted a survey of 1500 Americans about their attitudes towards history and the past. These authors concluded that professionally written and taught history had little resonance for their interviewees. The past, however, was very meaningful to the majority, so long as they could access it in tangible and personal ways, for example, by collecting antiques or memorabilia, preparing traditional foods for holidays, visiting museums and "living history" exhibitions, documenting their lives through photo albums (Rose, 2003), and pursuing family history research. African and Native Americans did tend to understand history more broadly in terms of their shared experiences of past injustice, while white American uses of the past were more family-oriented and personal. These findings caused Rosenzweig and Thelen (1998) to reflect upon the mismatch between their discipline's unappreciated conventional classrooms and publications versus the tremendous public appetite for a personal past. As geographers, we wonder how peoples' personal geographies of autobiography and family history might inform our academic practices.

While academicians can certainly belittle plebian entertainments such as tracking great-grandpa's birth certificate or displaying great-grandma's quilt (perhaps under the label of naïve "heritage", (*sensu* Lowenthal, 1996) scholars must guard against assuming that their own preferred leisure pursuits are thereby intellectually unassailable, and thereby dismiss the scholarly research potential of examining family history practices. Rather than criticizing family historians, perhaps geographers could direct them to useful land records, historical maps, or compendia of former place names (see Kashuba, Chapter 2 and Ruvane and Dobbs, Chapter 3, this volume), or suggest ways in which geographical insights on chain migration, residential segregation, or upward mobility might help to contextualize their ancestors' lives, as a form of professional service and community outreach (Hunter, this volume). Personal heritage tourism that is meaningful to family historians also has the potential to benefit host communities (Timothy, this volume). As DNA testing for genealogical purposes gains in popularity and sophistication, scholars can help family historians interpret those results (Mountain and Guelke, this volume) in credible ways. How genealogically relevant data are created, stored, and retrieved (see the chapters by Otterstrom, Ruvane and Dobbs, Meethan, and Richards in this volume) is another theme of common cause. Indeed, both scholars and private family historians may align on Hey's (1996: 2) rationale:

> Knowing one's ancestors is not a matter of mild curiosity: it is often part of an attempt to explain life and how we have come to be what we are, not just physically through inherited genes, but how we have come to believe in certain principles or to have acquired the attitudes, prejudices, and characteristics that mould our personalities.

Whether this pursuit is best undertaken through old letter collections in the attic, a GIS of land records, or through critical-theoretical expositions of power relations is a less important question than assuring that the process begins.

References

Adams, J.W. and Kasakoff, A.B. (1986), "Anthropology, Genealogy, and History: a Research Log", in Taylor, R.M, Jr. and Crandall, R.J. (eds), *Generations and Change: Genealogical Perspectives on Social Change*, Macon: Mercer University Press, pp. 53-78.

Allen, J.B., Embry, J.L. and Mehr, K.B. (1994-95), "Hearts Turned to the Fathers", *Brigham Young University Studies*, 34(2), 4-392.

Anderson, B. (1991), *Imagined Communities: Reflections on the Origins and Spread of Nationalism*, London: Verso Press.

Anderson, S.C. and Thode, E. (2000), *A Genealogist's Guide to Discovering Your Germanic Ancestors*, Cincinnati: Betterway Books.

Baker, R. (1995), "Life with Mother", in Zinsser, W. (ed.), *Inventing the Truth: The Art and Craft of Memoir*, Boston: Houghton Mifflin, pp. 23-38.

Basu, P. (2001), "Hunting Down Home: Reflections on Homeland and the Search for Identity in the Scottish Diaspora", in Bender, B. and Winter, M. (eds), *Contested Landscapes: Movement, Exile and Place*, Oxford: Berg, pp. 333-348.

Battaglia, D. (1999), "Towards an Ethics of the Open Subject: Writing Culture in Good Conscience", in Moore, H.L. (ed.), *Anthropological Theory Today*, Cambridge: Polity Press, pp. 114-150.

Bear, L. (2001), "Public Genealogies: Documents, Bodies and Nations in Anglo-Indian Railway Family Histories", *Contributions to Indian Sociology*, 35: 355-388.

Berry, K.A. and Henderson, M.L. (2002), "Introduction: Envisioning the Nexus between Geography and Ethnic and Racial Identity", in Berry, K.A. and Henderson, M.L. (eds), *Geographical Identities of Ethnic America: Race, Space, and Place*, Reno: University of Nevada Press, pp. 1-14.

Bhabha. H. (1994), *The Location of Culture*, London: Routledge.

Blickle, P. (2002), *Heimat: A Critical Theory of the German Idea of Homeland*, Rochester: Camden House.

Blunt, A. (2003), "Geographies of Diaspora and Mixed Descent: Anglo-Indians in India and Britain", *International Journal of Population Geography*, 9(4): 281-294.

Brault, G.J. (1986), *The French Canadian Heritage in New England*, Hanover: University of New England Press.

Brodwin, P. (2002), "Genetics, Identity, and the Anthropology of Essentialism", *Anthropological Quarterly*, 75: 323-330.

Brown, K. (2002), "Tangled Roots? Genetics Meets Genealogy", *Science*, 295(5560): 1634-1635.

Bruner, E. (1996), "Tourism in Ghana: The Representation of Slavery and the Return of the Black Diaspora", *American Anthropologist*, 98: 290-305.

Camp, A. (1996), "Family History", in Hey, D. (ed.), *The Oxford Companion to Local and Family History*, Oxford: Oxford University Press, pp. 168-174.

Carter-Smith, F. and Croom, E.A. (2003), *A Genealogist's Guide to Discovering Your African-American Ancestors*, Central Islip, NY: Betterway Books.

Canefe, N. (1998), "Citizens vs. Permanent Guests: Cultural Memory and Citizenship Laws in a Reunified Germany", *Citizenship Studies*, 2: 519-544.

Cavalli-Sforza, L., Menozzi, P. and Piazza, A. (1994), *The History and Geography of Human Genes*, Princeton: Princeton University Press.

Cobley, P. (2004), "Marketing the 'Glocal' in Narratives of National Identity", *Semiotica*, 150: 197-225.

Coles, T. and Timothy, D.J. (eds) (2004), *Tourism, Diasporas and Space*, London: Routledge.

Domosh, M. (2002), "Toward a More Fully Reciprocal Feminist Inquiry", *Acme*, 2: 107-111.
Duncan, J.S. (2006), "Cultural Geography", in Warf, B. (ed.), *Encyclopedia of Human Geography*, Thousand Oaks, CA: Sage, pp. 70-74.
Edensor, T. (2002), *National Identity, Popular Culture, and Everyday Life*, Oxford: Berg.
Elliott, C. (2003), "Adventures in the Gene Pool: DNA Testing and Identity", *The Wilson Quarterly*, 27: 12-21.
Embassy of Ireland, Washington, D.C. (2001), "Irish Citizenship by Descent (FBR)", <http:www/irelandemb.org/fbr.html>.
England, K. (1994), "Getting Personal: Reflexivity, Positionality, and Feminist Research", *The Professional Geographer*, 46: 80-89.
Finkler, K. (2001), "The Kin in the Gene: The Medicalization of Family and Kinship in American Society", *Current Anthropology*, 41: 235-264.
Fowler, S. (2001), "Our Genealogical Forebears", *History Today*, 51: 42-43.
Freeman, V. (2000), *Distant Relations: How My Ancestors Colonized North America*, Toronto: McClelland Stewart.
Friedensohn, D. and Rubin, B. (1984), "Generations of Women: A Search for Female Forebears", *History Workshop Journal*, 18: 160-169.
Fryxell, D.A. (2003a), "Simply the Best: 2003 *Family Tree Magazine* 101 Best Web Sites", *Family Tree Magazine*, 4(4): 20-31.
Fryxell, D.A. (2003b), "Decode Your DNA", *Family Tree Magazine*, 4(4): 66-67.
Giddens, A. (1990), *The Consequences of Modernity*, Stanford: Stanford University Press.
Giddens, A. (1991), *Modernity and Self-Identity*, Stanford: Stanford University Press.
Grubgeld, E. (1998/98), "Anglo-Irish Autobiography and the Genealogical Mandate", *Eire-Ireland*, 32(4)/33(1/2): 96-115.
Haley, A. (1976), *Roots: The Saga of an American Family*, New York: Dell Publishing.
Hasty, J. (2002), "Rites of Passage, Routes of Redemption: Emancipation Tourism and the Wealth of Culture", *Africa Today*, 49(3): 47-76.
Hey, D. (1993), *The Oxford Guide to Family History*, Oxford: Oxford University Press.
Hobsbawm, E. and Ranger, T. (eds) (1992), *The Invention of Tradition*, Cambridge: Cambridge University Press.
Huntington, E. and Ragsdale, M. (1935), *After Three Centuries: A Typical New England Family*, Baltimore: Williams and Wilkins.
Iossifides, M.A. (1991), "Sisters in Christ: Metaphors of Kinship among Greek Nuns", in Loisoz, P. and Papataxiarchis, E. (eds), *Gender and Kinship in Modern Greece*, Princeton: Princeton University Press, pp. 135-155.
Kahlke, M. and Kahlke, W. (1920), *Die Wappen der alten Bauernfamilien in den holsteinischen Elbmarschen*, Altona: Verlag von Riegel & Jensen (J. Harder, Buchhandlung).
Kashuba, M. (2005), *Walking with your Ancestors: A Genealogist's Guide to Using Maps and Geography*, Cincinnati, OH: Family Tree Books.

Lambert, R.D. (2002), "Reclaiming the Ancestral Past: Narrative, Rhetoric and the 'Convict Stain'", *Journal of Sociology*, 38: 111-127.

Louie, A. (2001), "Crafting Places through Mobility: Chinese-American 'Roots-Searching' in China", *Identities*, 8: 343-379.

Lowenthal, D. (1996), *Possessed by the Past: The Heritage Crusade and the Spoils of History*, New York: Free Press.

Martin, J.G. (1973), *Ellsworth Huntington: His Life and Thought*, Hamden: Archon Books.

Maiewskij-Hay, V. (1999), "Global History from the Local Perspective: An Instructional Technique", *Teaching History: A Journal of Methods*, 24(2): 71-77.

McCain, G. and Ray, N.M. (2003), "Legacy Tourism: The Search for Personal Meaning in Heritage Travel", *Tourism Management*, 24: 713-717.

McKee, J.O. (ed.) (2000), *Ethnicity in Contemporary America: A Geographical Appraisal*, 2nd ed., London: Rowman & Littlefield.

Mills, E.S. (2001), *Professional Genealogy: A Manual for Researchers, Writers, Editors, Lecturers, and Librarians*, Baltimore: Genealogical Publishing Company.

Molnar, S. (1998), *Human Variation: Races, Types and Ethnic Groups*, 4th ed., Upper Saddle River, NJ: Prentice Hall.

Monk, R.C. (ed.) (2002), *Taking Sides: Clashing Views on Controversial Issues in Race and Ethnicity*, 4th ed., Guilford: McGraw-Hill/Dushkin.

Morgan, N. and Pritchard, A. (2004), "Mae 'n Bryd I ddod Adref—It's Time to Come Home: Exploring the Contested Emotional Geographies of Wales", in Coles, T. and Timothy, D.J. (eds), *Tourism, Diasporas and Space*, London: Routledge, pp. 233-245.

Morris, H.F. (1992), "The 'Principal Inhabitants' of County Waterford in 1746", in Nolan, W. and Power, T.P. (eds), *Waterford History & Society*, Dublin: Geography Publications.

Moss, P. (2001), "Writing One's Life", in Moss, P. (ed.), *Placing Autobiography in Geography*, Syracuse: Syracuse University Press, pp. 1-21.

Nash, C. (2002), "Genealogical Identities", *Environment and Planning D: Society and Space*, 20: 27-52.

Nash, C. (2004), "Genetic Kinship", *Cultural Studies*, 18: 1-33.

Nash, C. (2005), "Geographies of Relatedness", *Transactions, Institute of British Geographers*, 30: 449-462.

Nash, C. (2006), "Irish Origins, Celtic Origins: Population Genetics, Cultural Politics", *Irish Studies Review*, 14: 11-37.

Ohlinger, G. (1919), *The German Conspiracy in American Education*, New York: George H. Doran Company.

Patterson, R.P. (1985), *The Seed of Sally Good'n: A Black Family of Arkansas 1833-1953*, Lexington: University of Kentucky Press.

Peach, C. (2000), "Discovering White Ethnicity and Parachuted Pluralism", *Progress in Human Geography*, 24: 620-26.

Ramirez, D. (2002), "Latin Lessons", *Family History Magazine*, 3(5): 54-61.

Rose, G. (2003), "Family Photographs and Domestic Spacings: A Case Study", *Transactions of the Institute of British Geographers*, 28: 5-18.

Rosenzweig, R. and Thelen, D. (1998), *The Presence of the Past: Popular Uses of History in American Life*, New York: Columbia University Press.

Rumbaut, R.G. and Portes, A. (eds) (2001), *Ethnicities: Children of Immigrants in America*, Berkeley: University of California Press.

Ryland, R.H. (1826) (reprinted 1982), *The History, Topography and Antiquities of the County and City of Waterford, with an Account of the Present State of the Peasantry of that Part of the South of Ireland*, London: John Murray.

Scott, J.W. (1992), "Experience", in Butler, J. and Scott, J.W. (eds), *Feminists Theorize the Political*, New York: Routledge, pp. 22-40.

Scott, J.C., Tehranian, J. and Mathias, J. (2002), "The Production of Legal Identities Proper to States: The Case of the Permanent Family Surname", *Comparative Studies in Society and History*, 44: 4-44.

Scupin, R. (ed.) (2003), *Race and Ethnicity: An Anthropological Focus on the United States and the World*, Upper Saddle River, NJ: Prentice Hall.

Shelton, J. (2001), "Lay My Burden of Southern History Down", *Southern Cultures*, 7: 100-103.

Sykes, B. (2001), *The Seven Daughters of Eve: The Science that Reveals Our Genetic Ancestry*, New York: W.W. Norton.

Taylor, R.M. and Crandall, R.J. (eds), (1986), *Generations and Change: Genealogical Perspectives on Social Change*, Macon: Mercer University Press.

Timothy, D.J. (1997), "Tourism and the Personal Heritage Experience", *Annals of Tourism Research*, 24: 751-54.

Timothy, D.J. and Teye, V.B. (2004), "American Children of the African Diaspora: Journeys to the Motherland", in Coles, T. and Timothy, D.J. (eds), *Tourism, Diasporas and Space,* London: Routledge, pp.111-123.

Tolzmann, D.H. (2000), *The German-American Experience*, Amherst, NY: Humanity Books.

Trommler, F. and McVeigh, J. (eds) (1985), *America and the Germans: An Assessment of a Three-Hundred Year History*, Philadelphia: University of Pennsylvania Press, 2 Vols.

Turnbaugh, W.A., Nelson, H., Jurmain, R. and Kilgore, L. (1993), *Understanding Physical Anthropology and Archaeology*, 5th ed., St. Paul: West Publishing Company.

Twyman, C., Morrison, J. and Sporton, D. (1999), "The Final Fifth: Autobiography, Reflexivity and Interpretation in Cross-Cultural Research", *Area*, 31: 313-325.

Waters, M.C. (2002), "The Social Construction of Race and Ethnicity: Some Examples from Demography", in Denton, N.A. and Tolnay, S.E. (eds), *American Diversity: A Demographic Challenge for the Twenty-first Century*, Albany: State University of New York Press, pp. 25-49.

Waters, M.C. (1990), *Ethnic Options: Choosing Identities in America*, Berkeley: University of California Press.

Watson J. (1996), "Ordering the Family: Genealogy as Autobiographical Pedigree", in Smith, S. and Watson, J. (eds), *Getting a Life: Everyday Uses of Autobiography*, Minneapolis: University of Minnesota Press, pp. 297-323.

White, R. (1998), *Remembering Ahanagran: Storytelling in a Family's Past*, New York: Hill and Wang.

Yang, P.Q. (2000), *Ethnic Studies: Issues and Approaches*, Albany: State University of New York Press.

Zelinski, W. (2001), *The Enigma of Ethnicity: Another American Dilemma*, Iowa City: University of Iowa Press.

PART I
TOOLS, SOURCES, AND IMPLICATIONS FOR GEOGRAPHY AND FAMILY HISTORY

Chapter 2

The Unfolding Tale of Using Maps in Genealogical Research

Melinda Kashuba

> A genealogist should be a composite—he should have the discrimination of a lawyer, the meticulous care of a mathematician, the sharp mind of a Sherlock Holmes, the patience of Job, the knowledge of an historian and a geographer, the imagination of a writer, and the pride of all men in a task well done. (Harland, 1963: 150)

Geography is not History

Sooner or later genealogists encounter the proverbial "brick wall" and discover the record trail that they have been following suddenly stops producing information on the individual or family line studied. The research question such as "Who is descended from whom?" suddenly shifts to "Where did so and so go?" Reliance on pedigree analysis, the study of family lineage and family groups with attention to primary and secondary source records, can only carry genealogical research so far in regions such as North America, Europe, and Australia given the high degree of mobility of their residents.

For example, the United States Census Bureau tells us that between 1995 and 2000, 46% of the U.S. population aged 5 and older (120 million) moved during that time (U.S. Census Bureau, 2003). Since 1790, the Census Bureau has been taking the pulse of the American population every ten years. At the time of the first census, it was no surprise that the mean center of the American population was nestled amidst the original thirteen colonies on the eastern shore of Chesapeake Bay in Kent County, Maryland. Subsequent decennial censuses depicted a mean center that hopped across Virginia, West Virginia, Ohio, Indiana, and Illinois. As shown in Figure 2.1, by the 2000 census, the center landed in south-central Missouri in Phelps County neatly demonstrating the peripatetic nature of the American population's westward movement.

With this sort of inclination toward movement, one would think American genealogists would be naturally interested in maps and other geographic tools and frequently make use of them as they track their wayward ancestors. Up until the decade of the 1990s nothing could have been less true. Before 1985, most American genealogists mirrored the American population by not having much general knowledge about geography or how to read maps.

There were many reasons for this situation but foremost was probably the languishing of geography in the American public education system. Often geography

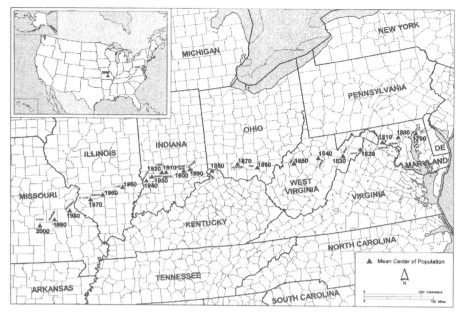

Figure 2.1   Mean center of population in the United States, 1790-2000
Source: United States Census Bureau <http://www.census.gov>

and map reading were subsumed in primary (ages 5-13) and secondary schools (ages 14-18) under "social studies" and combined (not identified as a separate discipline) with history and presented to students in bits and pieces rather than as a coherent subject of study. Since the mid-1990s the situation has begun to improve, with an articulation of national standards for geographic education in response to the "Goals 2000: Educate America Act" (Downs, 1995).

The American experience with flagging geographic education in the public schools is by no means confined to the United States. Other countries such as Canada and Australia have also reported an historic diminishment of geographic education in their public schools but that geography was again on the rise (Wolforth, 1986; Conolly, 2000). Great Britain, however, maintained a stronger geographic education in its public schools although it is allied closely with the study of history and the physical sciences. However, the time-worn cliché "history is about chaps, geography is about maps" seems to sum up much of the British public's view of geography after they left school.

American genealogists were also caught in a tradition that developed before World War I by a select few "professional genealogists" who worked primarily with New England records. New England is one of the most "records rich" regions in the United States because it was settled early by several groups of people from Europe who had the ability (literacy) and interest (partly due to religious ideology) to keep records. It is not uncommon for a genealogist pursuing ancestry in New England to have several hundred years of records to peruse. Maps were viewed by many early genealogists as "miscellaneous records" or ancillary to problem solving to be used only after the relative plethora of vital, civil, and church records was examined or to locate a place in relation to a civil or parish boundary.

These eastern genealogists *knew* maps existed but generally failed to see them as problem solving tools until much later. Instead the concentration of research effort was on the acquisition of other types of records. An example of the way maps were generally overlooked is *Index to Genealogical Periodicals* (Jacobus, 1963). This reference indexed many major American historical and genealogical periodicals of the day with a heavy emphasis on colonial genealogical subjects. "Maps" as a subject do not appear as a topic or are even subsumed as a subject in the locality index in either the original edition or the second edition (Boyer, 1988). Undoubtedly maps were used to illustrate points in some of these materials and the relevance of geography may have been discussed to support pedigree analysis but the lack of acknowledgement of maps, gazetteers, or geography as a subject relevant to genealogists' interests demonstrates how myopic the focus of genealogical research was at this time.

The publication of *The Researcher's Guide to American Genealogy* elevated the status of locality analysis by encouraging genealogists for the first time to "get them all"—gather, read, and extract information from all sources available including those related to localities such as maps, gazetteers, shipping guides, and postal guides (Greenwood, 1973). Greenwood emphasized that places, not just people, had genealogies. Greenwood believed that *thorough* genealogists paid attention to the innumerable boundary changes that affected the developing American nation. He opined that nearly every early American genealogical problem could be related back to boundary changes of one kind or another—"Even those families who lived in only one place had the boundaries changed around them, thus making it necessary to search the records of several counties or several towns" (Greenwood, 1973: 57).

Using Maps and Geography in Genealogical Research

One important step toward the dissemination of knowledge about genealogical research methods and sources (not just geographic ones) was the development of several national conferences and institutes designed to bring genealogists together to exchange ideas and learn from skilled professional and amateur practitioners. Genealogical conferences differ from other academic conferences because they trend toward instructional lectures and presentations rather than brief verbal reports or poster sessions based upon the results of on-going or completed research. Both the National Genealogy Society (NGS) and the Federation of Genealogical Societies (FGS) sponsored their first national conferences in 1978. Three years later, NGS inaugurated its annual "Conference in the States" which functions as its annual meeting and major educational event.

A perusal of the lecture topics presented at annual meetings between 1978 and 2004 demonstrates a growing interest and awareness of the use of maps and geography as sources and tools in problem definition and analysis. One of the earliest lectures presented at the NGS Diamond Jubilee Conference held in 1978 was by Ronald Grim, head of the reference and bibliography section of the Geography and Maps

Table 2.1 **A survey of the number of lectures presented at U.S. genealogical conferences specifically related to maps and geographic sources**

Time Period	1978-1979	1980-1989	1990-1999	2000-2004	Total
# of lectures concerning maps	2	26	102	47	177

Source: Joy Reisinger's index to papers given at the Federation of Genealogical Societies, National Genealogical Society and Gentech Conferences <http://www.fgs.org>

Division at the Library of Congress in Washington, D.C. (Grim, 1978). As Table 2.1 shows, the greatest growth in popularity of maps and geography-related subjects came during the decade of the 1990s.

Much of that growth can be attributed certainly to the popularity of the Internet as a source for both genealogical and geographical information including the availability of digitized historical map collections, online topographical maps, gazetteers, as well as an avenue for the dissemination of research methodology using these geographical tools. The number of national magazine publications devoted to genealogy also grew during the 1990s which gave more avenues for the dissemination of information about using maps in genealogical research. National periodicals that existed prior to 1990 also altered their formats by adding glossy, color graphics and user-friendly formats devoted to capture readers' interests that lent themselves to articles about maps. *Ancestry Magazine* and *Everton's Genealogical Helper Magazine* are two older magazine publications that attempted to increase their readership through format and editorial changes to compete against newer magazines such as *Family Chronicle* (founded in Canada in 1996), *Family Tree Magazine* (founded in the United States in 2000) and *Your Family History* (founded in the United Kingdom in 2003).

Two major national genealogical institutes have contributed to developing awareness and demonstrating the use of maps and geography as a means to solving genealogical problems. The annual week-long Institute on Genealogical and Historical Research (IGHR) and the National Institute on Genealogical Research (NIGR) are held respectively at Samford University in Birmingham, Alabama, and at the National Archives in Washington, D.C. NIGR routinely incorporates materials from the National Archives' Cartographic and Architectural Records Division into its curriculum.

In 1986 a layperson's guide to maps was published that stimulated the public's and genealogists' interest in different types of maps produced by government agencies and commercial publishers. *The Map Catalog* edited by Joel Makower not only described the dizzying array of thematic maps available but also attempted to classify them into categories easily comprehended by the general public (Makower, 1990). The book featured short sections and appendices on a variety of map-related topics including among many topics: copyright, map and atlas selection, map symbols, map software, lists of government sources, commercial publishers and map vendors, and college map library collections. Although over a decade out-of-print and out-of-date, Makower's work continues to be one of the most accessible guides on the subject of contemporary maps and is found on the reference shelves of many local public libraries throughout the United States.

Although Jacobus' tome was reprinted during the 1980s (Boyer, 1988), its revisions did not capture the efflorescence of genealogical writing that was taking place. In attempt to create a useful index to genealogical writings because many were not included in traditional academic social science indices, in 1985 the Allen County Public Library (Fort Wayne, Indiana) began publishing the *Periodical Source Index* (PERSI), an annual index to genealogical literature. Over the years, the Allen County Public Library developed a reputation for having one of the largest collections of genealogical periodicals in the nation. Its present index published online through agreement with HeritageQuest Online <http://www.heritagequest.com> and ProQuest Information and Learning <http://proquest.com/> indexes 10,000 periodicals published by local, regional, state, and national genealogical societies. Many larger public libraries subscribe to this database and offer it for free to their library card holders. The latest PERSI update released in October, 2006 included two million citations. In spite of the tremendous number of periodicals it includes, its index remains a little bit idiosyncratic but a search on the number of articles that include the words "map," "atlas," "boundary," "topographic," "geo," and "migration" produced 446 total articles as shown in Table 2.2.

Table 2.2 The number of articles published in genealogical periodicals related to maps and geography between 1950-September 2006 (based upon PERSI version 8.2)

Time period	1950-1969	1970-1979	1980-1989	1990-1999	2000-Sept 2006
# of articles	17	22	78	219	351

Note that the time period beginning in 1950 and ending in 1984 is the least accurate given that Allen County Public Library began its index in 1985 and is moving backwards in time slowly indexing material in its collection prior to the publication of its first annual index.

The titles of the articles represent the development of sophistication of the genealogical reader (and perhaps periodical editors). Prior to 1980, the articles were concerned with land platting, topographic maps and a variety of natural resource maps including soils maps to help enlighten the genealogist as to the qualities of their ancestor's physical landscape. During the 1980s there appeared to be great interest in using fire insurance maps, plat maps, city directory maps and many other types of maps to locate ancestral parcels as well as migration trail maps. The decade of the 1990s brought forth a plethora a "tips oriented" writing for genealogists including articles not only on the previously mentioned types of maps but also map symbols, terminology, map collections, and gazetteers. Since the year 2000, the same subjects continued to be written about but with more attention being paid to electronic sources of maps and online map collections.

One persistent thread that can be found both among the lectures presented at national meetings and articles printed from the 1960s through the year 2005 is "Why genealogists need maps and geography." The fact that this theme is repeated decade after decade reveals that genealogical researchers really do not know what geography and maps are all about and can not make a judgment how to use them effectively in their research. In 1998, a chapter dealing with geographic tools was published in the encyclopedic *Printed Sources, A Guide to Published Genealogical* Records which outlined the variety of atlases, gazetteers, governmental and commercially published map sources available (Schiffman, 1998). Within the past year two books have appeared to further illuminate the different types of resources available and the way maps can be applied to genealogical research: *Walking with Your Ancestors, A Genealogist's Guide to Using Maps* and *Geography and Genealogy, Geography and Maps: Using Atlases and Gazetteers to Find Your Family* (Kashuba, 2005; Douglas, 2006). Kashuba's title deals exclusively with American map sources while Douglas' book encompasses Canadian resources.

Why do Genealogists need Geographic Information?

Like everyone else, genealogists primarily consult maps to locate a place. Of course this only works if the place-name is identifiable and can be located on a modern or historic map. The process of acquiring the right map at the proper scale representing a time period similar to the time period the ancestor resided can be daunting to a neophyte researcher. Sources genealogists consult in order to locate maps are described later in this chapter.

The absolute location (latitude and longitude) of a place has acquired new significance in the genealogical community. Several popular genealogical database programs not only provide family historians with an opportunity to centralize their note keeping, pedigree charts and family group sheets and other documentation but also accept geographic coordinates such as latitude and longitude or coordinates obtained from global positioning system (GPS) devices. Images such as photographs and maps can also be attached to genealogical files within these programs. Personal genealogical websites are beginning to capitalize on this technology by using a combination of database information, images, and mapping technology. An example of this is the website for Kitterman Cemetery located in Dahlonaga Township, Wapello County, Iowa <http://soli.inav.net/~shepherd/kitt/kit_cem/cem1.html> which not only lists the information found on each tombstone but attaches a photograph of each grave marker, describes the condition, notes discrepancies from previous cemetery surveys and gives the geographic coordinates read from a GPS.

Genealogists also use maps to find out the distances and plan routes between places. This type of inquiry can be as mundane as asking "How far is my hotel from the archives?" and "What is the best way to get there?" But they also ask a map analytical questions such as "What is the distance my ancestor could travel in half a day's time by horse-drawn buggy to court a potential wife?" And by extension, "What are the settlements and farmsteads that fall within that time-distance sphere?" and "Can a line be inscribed on a map to represent an area served by particular routes that would help narrow a search for eligible females of courting age?"

Maps provide genealogists with the location of civil or religious jurisdictional boundaries. In the United States, records pertaining to marriage, property, and businesses are generally created and archived at the local jurisdictional level such as town, township or county. Although there is no guarantee that ancestors respected borders and found their way to the appropriate courthouse to register a deed or a vital record, knowing where the lines were drawn at a given time period is immensely valuable information that assists genealogists in focusing where to search for records as well as suggesting other adjacent jurisdictions that might have information.

Maps are useful tools to explore the relationship between a location on the map and its larger regional context. Characteristics of the physical environment of a region can suggest types of ancestral livelihood (and perhaps associated records). A homestead located near a waterbody, river crossing or road might suggest the connections to a transportation network to buy and sell goods, migrate or look for a spouse. It is possible that an ancestor might be prevented from traveling to the nearest courthouse due to obstacles such as a mountain range, seasonally difficult river crossing, or lack of reliable roads and opted to travel to a courthouse in another jurisdiction simply because it was more convenient or easier to reach. Moreover, genealogists study the larger area to gain an understanding of what factors might "push" an ancestor out of an area (drought, war, disease, thin soils, forced migration) and identify what factors might "pull" an ancestor into another area (better land, religious freedom, employment, mineral resources, familiar culture, family ties).

Maps are used to analyze residential location data collected by genealogists such as the presence or absence of families in a given county. The residential locations of a family or group of families can be traced backwards over time using sources such as the federal census and mapped showing how their lives weave together, separate and perhaps recombine in a new place following migration.

As a result several major software developers are betting upon genealogists' growing interest in the use of maps as analysis tools. For example, Progeny Software, Inc. <http://www.progenysoftware.com/> released *Map my Family Tree* as a way of cleverly combining location information from genealogical database programs and placing those locations onto base maps. The program enables users to view maps at state, national and worldwide scales showing locations derived from their genealogical database programs where vital events took place such as births, marriages and deaths. The press release for this program suggests that the maps can be used to track family migrations (Progeny Software, Inc., 2006). The program also features the ability to check genealogy data files for place-name "inaccuracies" based upon a gazetteer containing 3.3 million place-names that comes with the program.

Genealogists are taking advantage of Google Maps "mashups" and tools created for other industries such as real estate to depict and analyze location information. A mashup is a program that uses the free Google Maps program to create innovative ways to depict data. Programmers create map mashups that genealogists are using for family research. Three of the most popular free mapping tools are: MapBuilder <http://www.mapbuilder.net/>, CommunityWalk <http://www.communitywalk.com/>, and TripperMap <http://www.trippermap.com/>. Genealogists use these tools to create online maps that can be linked to their personal websites or downloaded as a web page and added to their own websites by obtaining a Google Maps API key. An example using MapBuilder to create a map to display genealogical data linked to specific towns is found at David Reeves' website which shows the birthplaces of his family in several English towns (Reeves, 2006). These tools allow for a limited degree of customization. For example, by changing the color of the marker symbol to signify a certain type of event (yellow for birth, blue for marriage etc.) or the color signifying a particular family surname, the distribution of events and family becomes readily apparent on a map. Information can be linked to each marker such as a photograph of the individual, a residence, or gravestone or a few lines of biographical statement. The map reader then can move the mouse from marker to marker and the additional data is displayed giving the viewer simultaneously a sense of both geography and history. To a younger and more computer literate generation, this type of genealogical presentation may be more inviting and fun to investigate.

Of the three tools mentioned above, MapBuilder appears to have added new features that enable the user to annotate each marker placed on a map. At the writing of this chapter it was Beta testing some of these new features such as automatically filling in latitude and longitude of each marker, an expanded description box that allows that marker to be linked to additional information found on a web page, a photograph, a complete street address and even tags used in indexing on the Internet.

The popular virtual earth viewing program, Google Earth, has caught the eye of genealogists because its user interface has become easier to use and its public version remains free of cost. A high-speed Internet connection and a newer computer (generally less than two to four years old) is all that is needed to be able to access its worldwide satellite imagery coverage. Improvements to placemarking that include the addition of photographic images such as a picture of an ancestor, an ancestral home, or tombstone have excited family history researchers with new possibilities of documenting their work and sharing it with other family members or with the Google Earth community. Google Earth can be used to create a personal family history tour (Hodges, 2005). *Map my Family Tree* and *Family Atlas* (created by RootsMagic Inc.) have the ability to geocode locations and upload them to Google Earth.

Locating Unfamiliar Place-Names

Most genealogists tend to reach for maps first rather than other geographic sources such as gazetteers or toponym dictionaries. This can often lead to frustration when the place is not readily found on a modern map. The problem is compounded with the genealogist's search takes them away from familiar jurisdictions and into unfamiliar territory that include new and strange sounding place-names that relate to past or present landscapes. These searches often end in bewilderment on the part of genealogist. William Lamble presented a map librarian's view of these frustrating forays by genealogists into map collections (Lamble, 1999). His advice to fellow map librarians was to encourage family historians to check gazetteers first and maps second in order to make sure that the correct place has been identified. He outlined several reasons for why a place-name is not readily found on a modern map giving examples from his experience with the genealogical community.

There are many reasons why a place cannot be located. One of the most common is that the place-name has been changed. Unlike other parts of the world such as

Europe, American place-names frequently change sometimes due to transfers of property ownership. A tavern, ferry crossing, mine or ranch can assume the moniker of the new owner and the old owner's name may pass from both memory and maps with startling suddenness. Towns were sometimes renamed to honor an individual or historic event, create a particular image more conducive to settlement, or even "sanitize" a bawdy past (Monmonier, 2006). The United States government through the office of the U.S. Board on Geographic Names worked hard in the late nineteenth and early twentieth centuries to standardize spellings and eliminate confusion in place-names. Along the way, some place-names were changed from the original native or European names to names easier to spell: place-names were simplified such as "La Fayette" became "Lafayette".

The way a place-name was recorded may also add to confusion. Not all early American clerks were well qualified in spelling, handwriting, or even sufficiently literate to serve as faithful recorders of documents such as land deeds, bills of sale, wills, or vital records. Genealogists are well aware of the variations of surname spelling but that knowledge does not always translate into the arena of place-name searching. The variety of languages and accented English spoken on the western frontier especially in areas experiencing mineral rushes added to the problem of transcription of place-names. For example, a foreign-accented pronunciation of "Oakland" (Alameda County, California) easily morphed into "Ogden" and then was mistakenly transcribed by a clerk who misunderstood what was being said.

Another very common problem is the multiple occurrences of certain popular place-names in many locations. As people migrated, they often copied their place-names in their new locations. The geographer John Leighly noted forty-seven towns, counties or communities in the western half of the United States named "Salem" (Leighly, 1978). How conscious genealogists are of the pervasiveness of "creep" or diffusion of place-names across the landscape along with their migrating ancestors is unknown. The USGS notified the world that the most common place name in the United States was "Mill Creek" with 1,473 occurrences on American topographic maps (USGS, 1996). Canada, not to be outdone, let the world know that "Mount Pleasant" was its most common place-name and "Long Lake" its most common topographical name (Canadian Geographical Names, 2006).

One of the easiest solutions to the dilemma of finding an unfamiliar place-name is to locate a gazetteer compiled around the time period of the document containing the place-name in question. Because the United States grew so quickly following the American Revolution, the United States Post Office as well as commercial shippers of goods became involved with the publication of a variety of finding aids such as shipping guides or postal guides that listed towns by county and state and noted the availability of railroad stations, post offices, and telegraph offices. One of the largest producers of atlases and maps, the Rand McNally Company, based in Chicago, also printed maps, tickets, brochures, and itineraries for the many railroad companies. For example, a private genealogical library has placed on its website a copy of Rand McNally's *The New 11 x 14 Atlas of the World* published in 1895 and the *1891 Grain Dealers and Shippers Gazetteer* (Mardos Memorial Library, 2006). Both sources list cities, towns and small villages served by railroad companies during the late nineteenth century. Since 1869, the Rand McNally Company published a version of its popular annual *Commercial Atlas and Marketing Guide*. This atlas has always featured a large gazetteer section. The present edition is published in two volumes and lists 120,000 place-names as well as showing civil jurisdictional boundaries such as townships. Most public libraries and Family History Centers (local genealogical libraries sponsored by the Church of Jesus Christ of Latter-day Saints) have at least one copy of an earlier edition of this atlas.

For locating gazetteers from other countries that are either available as reproductions in print or on CD-ROM or available online, genealogists often turn to the genealogy portal known as "Cyndi's List" <http://www.cyndislist.com>. Over the past eleven years, Cyndi Howell has categorized and linked over 263,000 websites. She includes a category entitled "Maps, Gazetteers, and Geographical Information" for nearly every nation. The very popular geographic portal "Odden's Bookmarks—The Fascinating World of Maps and Mapping" <http://oddens.geog.uu.nl/index.php> hosted by a university in the Netherlands also provides links to online gazetteers and geographical dictionaries throughout the world. Many of the older place-name resources are beyond copyright and have been placed online as digitized scans. The GENUKI website for genealogical research in the United Kingdom and Ireland <http://www.genuki.org.uk/> features a virtual reference library of historical materials that includes both old gazetteers and historical maps.

The Family History Library located in Salt Lake City, Utah, which is operated by the Church of Jesus Christ of Latter-day Saints, loans microform versions of old gazetteers and geographical dictionaries to local Family History Centers throughout the world. The library catalog is available online at <http://www.familysearch.org>. The Genealogical Society of Utah which works in conjunction with the Family History Library microfilms geographic materials throughout the world with an eye toward preservation and sharing of materials of interest to amateur and professional researchers. The Family History Library website features research guides to geographical areas which are accessible by clicking on the "search tab" and then the "research helps". An alphabetical index is available to the searcher to make a selection. The research guides are abbreviated descriptions of resources available in the library and on the Internet pertaining to a variety of subjects and geographic locations. An example of one such guide describes the Meyers Gazetteer of the German Empire published in 1871 (*Meyers Orts-und Verkehrs-Lexikon des Deutschen Reichs*). This geographical reference has been microfilmed and due to its popularity is available in most Family History Center libraries as part of their standard reference collection.

Government agencies are also favorite haunts for genealogists to locate past and present place-names. The United States Geological Survey (USGS) was on the forefront of collecting place-names when it began its mission of resource mapping the entire nation in 1879. USGS began its program six years after the Ordinance Survey completed its mapping of England. The United States did not complete its topographic mapping of the entire country until prior to the decennial census of 1990.

USGS has published in one form or another (paper, microform, online) a version of its place-name database for many years. The online version known as the Geographic Names Information System (GNIS) <http://geonames.usgs.gov> allows for searches for every place in the United States, its territories and even Antarctica) that is listed on a topographical map or other federal documents. The database contains nearly two million entries including the names of places that no longer exist and variant names for existing places. Figure 2.2 shows a sample query result for the place-name "Heaven." The list includes everything from schools (Goose Heaven School) and cemeteries (Heaven City Burial Grounds) to river rapids ("Heaven Help You Rapids") and populated places (Heaven Heights). The results link to a table that gives a precise geographic description with coordinates and citation where the place-name appears in federal records. The table links to several choices of how the place can be displayed such as the National Map Viewer <http:// http://nmviewogc.cr.usgs.gov/viewer.htm> which generates a location map using a USGS topographic map, Google Maps <http://maps.google.com> including its map, aerial photograph and "hybrid" versions that list both marked locations and street names on an aerial photograph base map and TerraFly <http://terrafly.com> aerial photography at several scales of detail.

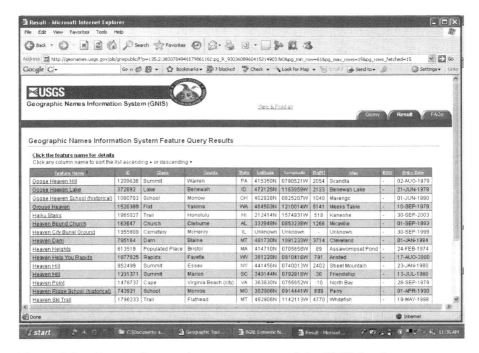

Figure 2.2 Results from a place-name query of the GNIS database

Source: U.S. Geological Survey, viewed 14 October 2006, <http://geonames.usgs.gov>

USGS realizes the importance of the genealogical community and has been an active participant at many national and regional genealogical conferences both in training genealogists to use their maps and displaying new geographic tools created for the general public. As an indicator of their early interest in the genealogical community in the 1980s, USGS published a fact sheet entitled "Using Maps in Genealogy"

(USGS, 2002). USGS released a brochure summarizing the maps, aerial photographs and databases of particular interest to genealogists (USGS, 2006). The timing of this brochure is crucial to the dissemination of information about changes in USGS printing of its popular 7.5 minute quadrangle series (1:24,000 scale), the ongoing digitization of maps, the availability of historic maps and aerial photographs from USGS, from both the National Archives and the Library of Congress, and the GNIS database.

Land and Property Maps

Genealogists often find land records particularly important because often they pre-date other records (with the exception of marriage records because they had a bearing on property inheritance and legitimacy issues) for a jurisdiction. A common research technique employed by genealogical researchers is to use indirect evidence gleaned from records such as land deeds to construct a network of family relationships when direct evidence is lacking. The mapping of property parcels based upon deed information can help create a picture of relationships when other records are absent. It is not unusual to see family members including in-laws living in close proximity to one another. Maps can help demonstrate these family connections.

In the United States there are two main types of land partitioning. One arrived from Europe with immigrants from countries such as Spain, France, and Great Britain known as the *metes and bounds* or *indiscriminate-survey system*. The latter phrase is sometimes applied because the land was chosen indiscriminately (independently) of a survey system. The original thirteen colonies including the derivative states of Maine, Vermont, Tennessee, Kentucky and West Virginia plus Hawaii and Texas (the last two never used the federal system) followed this system because a head-of-state granted land ownership to the colony or major proprietor(s), who then distributed the land to other individuals. These states became known as *state-land states* because unclaimed land at the time of the American Revolution or at the time of their transfer to the United States remained with the state rather than transferred to the federal government. States that were primarily formed from land owned by the federal government were known as *public-land states*.

The metes and bounds land description system relied upon the ability of surveyors to describe property boundaries and corners according cardinal directions, distance measurements, and to the arrangement of permanent natural features (such as landforms and streams), the boundaries of adjacent owners' properties, and ephemeral features such as trees or piles of rocks set as property corners. Early American land owners even practiced the ancient European custom of annual walks along the property lines (known as "beating the bounds") but for secular reasons rather than to follow religious customs. As this custom faded from practice so also went faith in the metes and bound system.

The other land survey system was born out of frustration with the metes and bounds system with the assistance of Thomas Jefferson before his election to the presidency. Jefferson proposed a national surveying and mapping program to inventory western lands prior to their sale to the public to help pay the national debt generated by the American Revolution. The Land Ordinance of 1785 and the Northwest Ordinance of 1787 provided impetus for the creation of the U.S. Public Land Survey (USPLS) that enabled surveying and settlement of lands ceded by the original thirteen colonies to the federal government as well as lands later acquired from Spain, France, and other countries. The USPLS utilized the "township and range" survey scheme which amounts to a grid of north-south trending meridians and east-west coursing baselines across the United States for the purpose of locating any parcel of land, no matter how small, prior to public sale.

The transfer of public land to private parties generated records of great interest to genealogists casting around for their western-migrating ancestors. These records were maintained by the General Land Office (GLO) which was formed in 1812 and later was merged in 1946 with another federal agency to form the Bureau of Land Management (BLM). The historic records of the GLO and BLM are retained by the National Archives and Records Administration (NARA) in Washington, D.C. and the regional NARA branches located throughout the United States.

The Homestead Act of 1862 and the General Mining Law of 1872 assisted in the settlement of land in the western United States by making additional lands available for purchase and refining the means by which federal lands could be transferred to private ownership including lands that had been settled or mined illegally. The BLM is in the process of posting online images of federal land title records (1820-mid 1960s), land surveys (dating back to 1810), and survey plats on its website <http://www.glorecords.blm.gov>.

Survey plats offer researchers another means to locate exactly where their ancestor purchased lands or may have squatted illegally. Once land was transferred from the federal government by way of a patent, subsequent land transfers are usually found in deed records of the county recorder or clerk (or township, municipality or civil parish records depending upon the state).

Land records, maps, and surveys related to the state-land states are found in state land offices and local recorder and clerk offices. Land grants made by foreign governments to individuals in portions of what was to become the United States (for example California, Arizona, New Mexico, Florida, Indiana, and Louisiana) often became the subject of land claims disputes in federal courts after these areas became states. State and federal archives often are repositories for these records and maps.

A series of county map books based upon GLO land patent records is being published and marketed by Arphax Publishing Company <www.arphax.com>. The production of these maps relies on creating map layers based GLO survey plat maps, initial land patentee lists, and the locations of cemeteries, water courses, and transportation networks. The work is an innovative application of geographical information systems (GIS) to locating cultural features and property lines. Problems regarding scaling of maps and exact locations of properties and roads are briefly noted by the publisher in each volume. The series is entitled *Family Maps of __ (name of county, state)*. The publisher is gradually working its way through all of the public-land states. Prior to June 2006 the publisher featured two types of map books—one that only focused on mapping homestead and a "deluxe edition" that included homesteads as well as the cultural and drainage features previously mentioned. Since June, the publisher has produced deluxe editions only for each county because of the popularity of the deluxe edition that features modern road and railroad networks superimposed upon a base map featuring historic boundary lines and land parcels.

Arphax is not the first commercial publisher to create a market for a popularized version of land plats. During the nineteenth century and early twentieth century commercial publishers created a special type of cadastral mapping unique to United States and Canada known as county maps or county atlases which sometimes are referred to as "plat books" although they were not produced by the federal government but were based upon federal land office records, county recorder's offices, and interviews of local residents by publisher representatives.

The maps and county atlases were popular in rural areas because they inventoried the distribution of land and showed the names of individual property ownership on a parcel by parcel basis. Nineteenth-century merchants delighted at being able to identify their customers, shippers were able to locate farms and homesteads, and land owners took particular pride in being able to locate their properties on maps. The maps held special appeal to land owners and these maps were aggressively marketed to them because they were produced on a subscription basis and for the publisher to make money, they had to sell many map books. Genealogists find these maps immensely valuable because of the placement of individual landholder's names on parcels. Cultural features such as orchards, cemeteries, schools, and churches were also frequently noted suggesting other record locations to research.

Between 1850 and the 1920s county cadastral maps, county atlases, and plat books were produced. The "golden age" of production occurred between 1850 and 1880 when the most detailed maps were created. Gradually, the maps and atlases were stripped of many of the more ornate and decorative features only to emerge as more utilitarian references by World War I (Conzen, 1990). Rural county plat books and directories are published today by several regional and national companies such as Farm and Home Publishers (Belmond, Iowa) and Rockford Map Publishers (Belvidere, Illinois) and used by genealogists in conjunction with modern road atlases. The largest collections of these atlases are held by the Library of Congress in Washington, D.C. and The Newberry Library in Chicago, Illinois.

County atlases were also published for Canadian counties by subscription. The website "In Search of Your Canadian Past: The Canadian County Atlas Digital Project" sponsored by McGill University <http://digital.library.mcgill.ca/countyatlas/> has uploaded a large number of Ontario, Quebec, and Maritime county atlases published between 1874 and 1881. The project sponsors plan to expand the online collection to cover counties in other provinces. Unlike other American county atlases that have been digitized and made available through the Internet, this project features a searchable database of the property owner's names that appear on the township maps in the county atlases. All maps, sketches of properties and portraits have been scanned and indexed.

Another type of parcel map genealogists frequently consult when researching urban ancestors or house histories are fire insurance maps. Unlike county atlases and maps, fire insurance maps generally did not show property ownership or the names of individual property owners unless they are linked to the name of a particular building or block. Fire insurance maps according to Diane Oswald show "the footprints of

America's industrial revolution" because they documented the rapid, post-Civil War growth of urban areas in the United States (Oswald, 1997).

Fire insurance maps were developed during the eighteenth century in Great Britain, the seat of the European industrial revolution, as a reference tool for insurance underwriters. Against the backdrop of burgeoning cities and their industrial cores on both sides of the Atlantic, the insurance industry began to grow and insure buildings further away from their headquarters. It became costly and inconvenient to dispatch agents to check the premises of potential insurance clients and assess possible risks before rendered a judgment on whether or not to insure a property. Entrepreneurial engineers and map publishers created detailed plans that showed physical characteristics of individual buildings such as type of roof material, number of rooms, how far it was from the nearest hydrant, width of streets, and adjacent land uses.

Genealogists and many other urban researchers find these maps useful because when combined with other sources such as deeds, city directories, and federal census records, they become a powerful reference source when one attempts to reconstruct past landscapes, study employment opportunities, search for nearby churches (which may hold records of interest), or identify neighborhood schools (another potential record source). All buildings were mapped—barns, outhouses, garages, gazebos, and porches were delineated giving insight into economic status and lifestyle of the person who dwelt on residential parcels. These intimate details provide information about ancestors not readily available elsewhere. The maps were updated fairly often, usually every five years or less depending upon the amount of development that had taken place between reconnaissance trips.

In the United States, the Sanborn Map Company became the major fire insurance publisher. Founded in 1867 in New York City, the location of several horrendous city fires, the Sanborn Company grew steadily merging with several competitors until by 1924 it mapped 11,000 communities throughout the United States. Gradually, very old Sanborn maps (pre-1923) are finding their way online as part of state and regional digital map collections because they are no longer under copyright restrictions. Environmental Data Resources, Inc. (EDR) purchased the library of historical maps and rights from the Sanborn Map Company. During the 1980s, Chadwyck-Healy, which is today part of ProQuest Information and Learning, microfilmed the Library of Congress collection. The black and white images on the microfilm have since been licensed by EDR and digitized by ProQuest. Access to this database is sold on a subscription basis to libraries and institutions (Kashuba, 2006).

The British Library <http://www.bl.uk/collections/maps.html> houses an extensive collection of fire insurance maps published in London by Charles E. Goad, Ltd. between 1885 and 1975. Charles E. Goad also produced fire insurance maps of Canadian towns and cities. A national inventory of Canadian fire insurance maps has been published (Dubreuil and Woods, 2002).

Border Blues

As was stated by Greenwood (1973) earlier in this chapter, most problems encountered by genealogists in the United States involve failing to perceive the importance of border changes. An understanding of how borders fluctuated through additions, splits, mergers, and occasional "extinctions" of counties is paramount to locating records. Identifying at what time period each county boundary change occurred may suggest what jurisdiction holds records. Not all records were transferred if a county were split in half and a new county created. The "parent county" often retained records created during the time period prior to the split. Inattention to such details is one of the most frequent causes for unsuccessful genealogical searches.

One key American reference genealogists look to is *Map Guide to the U.S. Federal Censuses, 1790-1920* (Thorndale and Dollarhide, 1987). The maps feature statewide county configurations keyed to the federal census decades. Notes added to each map explain changes that occurred between census years. The maps depict modern county configurations as white lines and historical county boundary lines as black lines. The map reader can immediately observe during any given census year the changes in county shapes and sizes compared to the present situation.

The Atlas of Historical County Boundaries Project is a project sponsored by the William M. Scholl Center for Family and Community History at the Newberry Library in Chicago. The project has published volumes documenting every boundary change that occurred to each county, unorganized territory, and non-county areas (that were attached to organized counties for administration purposes) in a given state. Each change is illustrated on a map and documented in a tabular format. A respective volume includes an index that captures every place-name that appears on the county maps.

Since 2001, the Atlas of Historical County Boundaries Project has shifted away from producing printed atlases toward the production of cartographic files that could be used in GIS systems. The maps are created in a digital format and are accessible through the project's website at <http://www.newberry.org/ahcbp/sitefiles/state_index.htm>.

A slightly different approach to boundary mapping and documentation is being applied to the English Jurisdictions Mapping Project (EJMP) sponsored by the Family History Library in Salt Lake City, Utah. The EJMP takes boundary mapping a step further by embedding consolidated data such as library catalog entries, indexes, record location information, relevant local histories, published family histories, and other genealogically important finding aids into the map using GIS. At the writing of this chapter, a seven-county prototype program has been completed with a goal toward extending the EJMP to the remaining thirty-three counties of England (Young, 2005).

The Association of British Counties <http://www.gazetteer.co.uk> and GENUKI "United Kingdom and Ireland Genealogy" <http://www.genuki.org.uk/> advise those seeking to understand the historical origins of the various types of administrative regions that are found in the British Isles. Although neither website documents individual boundary changes, both feature detailed explanations of how British counties differ from the concept of counties in other parts of the world.

Perhaps nowhere in the world are boundary changes more complicated than in the region of Eastern Europe, Germany, and the former Soviet Union. Genealogists pursing ancestry in this part of the world are frequently thwarted by national boundary and place-name changes. The Federation of Eastern European Family History Societies (FEEFHS) was organized in 1992 to facilitate research in eastern and central Europe. The organization's website <http://feefhs.org/> contains an extensive collection of digitized historical maps covering this region. The online map room is organized by: Austro-European Empire; German Empire East; German Empire West; Balkans; European Russia Empire; Asian Russia Empire; Harmsworth Atlas; Hutterite Map Collection; and Scandinavia (Finland).

The Perry-Castañeda Library Map Collection at the University of Texas <http://www.lib.utexas.edu/maps/> is creating an extensive digitized collection of modern and historical maps and atlases from all over the world. At this writing, they have uploaded 11,000 maps from their collection and made them available for free to the public. Their website links to other digital map collections throughout the world enabling the researcher to search for maps of a particular theme or location.

Besides large map collections and portals discussed previously, the WorldCat network of library materials and services <http://www.worldcat.org/> has cataloged one billion items including maps and atlases in over 10,000 libraries worldwide. Since WorldCat was released in a BETA version to allow access by any computer connected to the Internet, it has grown in popularity with genealogists. WorldCat can list search results according to the physical location (by zip code) of the searcher ranking hit results from the closest institution to the most distant.

Consumers become Producers of Maps

Maps have never been more available to genealogists than they are today through the Internet and in print. Recently the Geography and Map Division at the Library of Congress <www.loc.gov/rr/geogmap> announced that it placed its 10,000th map online at its website (Library of Congress, 2006). This noteworthy milestone was reached ten years after the library began digitizing maps from its collection. The World Wide Web has assisted genealogists in becoming more connected as a research community through mail lists, bulletin boards, websites, e-zines, blogs, skypecasts, and podcasts. Genealogists exchange information about sources and research techniques in all of these arenas besides traditional print media, regional and national conferences, and institutes.

As was demonstrated earlier in this article, the interest in maps as measured by the number of lectures presented at national conferences and articles appearing in genealogical publications has been growing steadily since the 1970s. Genealogists are learning that maps can enrich the information they find in other sources. Surrounded by a fast-paced, 21st century culture that is "on the move," being able to place one's finger on a spot on a globe or on a map and say "I come from here" or "my people used to live here" and "this is what I have found out about that place" from a gazetteer or a map takes on a certain poignancy. Placing aside novelist Tom Wolfe's title taken from an old maxim, "you can't go home again," through an investigation of historical maps and other geographic resources a genealogist can get closer to the past than ever before.

The geographic dimension of an ancestor's life fleshes out the names, dates, and places listed on pedigree lineage forms or family group sheets. An ancestor's migration can be viewed in the context of others migrating during the same time period under similar circumstances allows the researcher to speculate on motives, decision-making, and passions that drove people of the past to make certain choices. Knowledge of how one's ancestors got through the trials and tribulations of life related to where they were born, schooled, worked, worshipped, married, bred, and died can give context to the way we live our lives today. We, like our landscapes, are the product of those who came before us. The unfolding tale of how maps and geography can inform and enrich our lives continues. Through the ability of online resources such as Google Earth, Community Walk, and genealogical software, genealogists are quickly moving from consumers of geographic information or uncritical readers of maps to producers of maps keyed to the important dates and places related to their ancestors' lives. As geographers we can either ignore their work as the humble diversions of amateurs or we can welcome eager and inexperienced students who desire to learn from us about map history, diversity, and principles of construction that will lead them to think critically about geographic information and enhance the use of maps in their research pursuits.

References

Boyer III, C. (1988), *Index to Genealogical Periodicals*, Newhall, CA: C. Boyer.

Canadian Geographical Names [website], accessed December 2006 <http://geonames.nrcan.gc.ca/>

Conolly, G. (2000), "Promotion of Geography in Australia: An Unfinished Story", *International Research in Geographical and Environmental Education*, 9(2): 160-165.

Conzen, M. (1990), "North America County Maps and Atlases", in D. Buissert (ed.), *From Sea Charts to Satellite Images: Interpreting North American History through Maps*, Chicago: University of Chicago Press, pp. 186-212.

Douglas, A. (2006), *Genealogy, Geography, and Maps: Using Atlases and Gazetteers to Find Your Family*, Toronto: The Ontario Genealogical Society.

Downs, R. (1995), "Geography for Life—The New Standards". *UPDATE, the Newsletter of the National Geographic Society's Geographic Education Program*, [website] accessed 15 October 2006 <www.nationalgeographic.com/education/standards.html>

Dubreuil, L. and Woods, C. (2002), *Catalogue of Canadian Fire Insurance Plans 1875-1975*. Occasional Papers No. 6, Ottawa: Association of Canadian Map Libraries and Archives.

Greenwood, V. (1973), *The Researcher's Guide to American Genealogy*, Baltimore, MD: Genealogical Publishing Company.

Grim, R. (1978), "Maps, Surveys and Land Records", Paper presented at the National Society of Genealogy's Diamond Jubilee Conference, Silver Spring, Maryland.

Harland, D. (1963), *Genealogical Research Standards*, Salt Lake City, UT: Bookcraft.

Hodges, G. (2005), "Intro to Google Earth", (published online 17 August 2005) <http://apoetsblues.typepad.com/studentsofdescent/2005/08/intro_to_google.html> (blog), accessed 1 March 2007.

Jacobus, D.L. (1963), *Index to Genealogical Periodicals*. Vol. 1, Baltimore: Genealogical Publishing Company.

Kashuba, M. (2005), *Walking with Your Ancestors, A Genealogist's Guide to Using Maps and Geography*, Cincinnati, OH: Family Tree Books.

Kashuba, M. (2006), "Turn Up the Heat with Fire Insurance Maps", *NGS NewsMagazine*, 32(2): 26-29.

Lamble, W. (1999), "Genealogical Geography: Place Identification in the Map Library", 65th Annual Conference of the International Federation of Library Associations and Institutions (IFLA) [website], <http://.ifla.org/IV/ifla65/papers/045-94e.htm> accessed 13 October 2006.

Leighly, J. (1978), "Town Names of Colonial New England in the West", *Annals of the Association of American Geographers* 68(2): 233-248.

Library of Congress. "Geography and Map Division Announces Milestone of 10,000th Map Placed Online,' press release 06-183, published online 27 September 2006, [website] <http://www.loc.gov/today/pr/2006/06-183.html>

Makower, J. (ed.) (1990), *The Map Catalog*, 2nd edition, New York: Tilden Press.

Mardos Memorial Library (2006), Pam Mardos Rietsch, [website] <www.memoriallibrary.com>, accessed 13 October 2006.

Monmonier, Mark (2006), *From Squaw Tit to Whorehouse Meadow: How Maps Name, Claim and Inflame*, Chicago: University of Chicago Press.

Oswald, D. (1997), *Fire Insurance Maps: Their History and Applications*, College Station, TX: Lacewing Press.

Progeny Software Inc. (2006), [website] <http://www.progenygenealogy.com/map-my-family-tree.html>, accessed 11 October 2006.

Reeves, D. "Ancestors – Genealogy with Web 2.0.' [website] <http://home.earthlink.net/~dcreeves2000/data/gen_map_dr_web.htm>, accessed 12 October 2006.

Schiffman, C.M. (1998), "Geographic Tools: Maps, Atlases, and Gazetteers", in K.L. Meyerink (ed.) *Printed Sources, A Guide to Published Genealogical Records*, Salt Lake City: Ancestry Inc., pp. 95-146.

Thorndale, W. and Dollarhide, W. (1987), *Map Guide to the U.S. Federal Censuses, 1790-1920*, Baltimore: Genealogical Publishing Company Inc.

U.S. Census Bureau (2003), "Geographical Mobility 1995-2000, Census 2000 Brief", [website], <http://www.census.gov/prod/2003pubs/c2kbr-28.pdf> accessed 29 Sept 2006.

USGS (1996), "Latest Score: Mill Creek 1,473, Spring Creek 1,312", published online 3 September 1996, [website], <http://www.usgs.gov/newsroom/article.asp?ID=777> accessed 13 October 2006.

USGS (2002), "Using Maps in Genealogy", Washington, DC: USGS. PDF version [website] <http://mac.usgs.gov/isb/pubs/factsheets/fs09902.html>, accessed 14 October 2006.

USGS (2006), "Using USGS Resources for Research in Genealogy", PDF version [website] <http://education.usgs.gov/common/resources/genealogy_usgs.pdf>, accessed 14 October 2006.

Wolforth, J. (1986), "School Geography—Alive and Well in Canada?", *Annals of the Association of American Geographers* 76(1): 17-24.

Young, S.C. (2005), "The English Jurisdictions Mapping Project", in *World Library and Information Congress: 71st IFLA General Conference and Council Conference Programme*, [website] <http://www.ifla.org/IV/ifla71/Programme.htm>, accessed 13 October 2006.

Chapter 3

Genealogy, Historical Geography, and GIS: Parcel Mapping, Information Synergies, and Collaborative Opportunities

Mary B. Ruvane and G. Rebecca Dobbs

Introduction

A genealogist's ultimate goal is to identify and verify each relation along his or her family tree. An experienced family historian knows the importance of documenting not only direct descendants but members of the extended family, the people they interacted with, and the critical events and places that shaped and defined the lives of each generation. Only a careful compilation of information will facilitate the interpretation of evidence discovered and authenticate a family's history and the events that shaped them.

Critical to a genealogical investigation is determining the locations of past ancestral habitats and translating them into present day geographic terms. Knowing where and when a relative lived in the past and how ancestral places are administered today is instrumental in the hunt for relevant documents, as official records likely to contain needed clues may have been transferred to new jurisdictions coincident with shifts in administrative boundaries. Identifying the precise historical location where an ancestor lived is especially important in cases involving relatives that settled near a former county boundary, since more than one controlling authority may have inherited records useful in retracing their footsteps.

Maps are important tools for understanding past landscapes. They help a researcher to visualize the relationship between cities, towns, and county boundaries that existed during an ancestor's lifetime. The importance of maps to genealogical investigations is highlighted in numerous research guides on the subject, exemplified in works by Greenwood (2000: 59-60, 92-3), Helmbold (1976: 80-81), Kashuba (this volume), and the U.S. Geological Survey. Not only are maps beneficial for pointing to the possible location of archival records, but for understanding the influence of surrounding geographic features. For example, early topographic maps illustrate transportation routes and physical barriers such as mountain passes and waterways that may have limited interaction between communities (Maptech, 2007). Highly coveted are rare larger-scale early historical maps, such as land ownership and town street maps.

While most genealogists and local historians lack the skills for presenting their findings within a GIS, or geographic information system, many have employed graphic methods to visualize the locale of interest. Hand rendered sketches are a common technique, according to genealogical publications (Baugh, 2004: 30; Kashuba, 2005: 27). A prescient multimedia example is shown in a 1935 sketch and accompanying photographs (see Figure 3.1), found in a family archive, that illustrate how "the old neighborhood" had changed since the family sold their residence nearly 35 years earlier. Two local North Carolinians employed unique methods. One created hundreds of paper cut-outs representing early nineteenth century parcels to piece together his neighborhood of interest (G. Thomas, personal communication, February 8, 2005), the other, a retired surveyor, is rendering 1:24,000 scale maps of the community where his wife's family once resided (R. Booth, personal communication, February 16, 2005). Another method is to position the names of early settlers in the approximate location of the land they settled using an existing map for background. Still others have adopted the use of simple parcel mapping software (Hill, 2006). Although most of the resulting visual representations are primitive and not dimensionally precise, they usually do portray the relative position of neighboring land and other important cultural features.

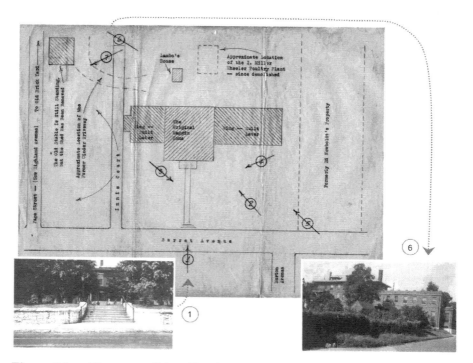

Figure 3.1   Changes to "the old neighbourhood" in Louisville, Kentucky

Source: Ruvane/Stewart Family Papers. Letter (8p), map, 12 photos. From Stewart Taylor, Louisville, KY, to Uncle Lou (Stewart), [NYC or Pocono Manor, PA]. 21 July 1935.

Yet GIS is a potential tool of enormous scope, one that would aid not only genealogists but also historians and historical geographers. In the United States, each group might welcome the possibility of creating a national map, using a GIS interface that chronologically depicts settlement patterns in America from the colonial period forward, ultimately linking to present-day digital county cadastral records. Another multi-user technology would be a multimedia Internet-accessible tool that not only enables researchers visually to explore changes in land ownership, jurisdictional boundaries, and related cultural features over time, but one where images of primary source material and supporting data could be viewed, shared, expanded upon, and enhanced as contributed research findings become linked. A network hosted by a consortium of libraries, archives, museums, and historical societies, acting as the central community hub, perhaps would reap the greatest benefit. As researchers' contributions grew their input could improve access to scattered but related institutional and private holdings, aided by new links between resources and an improved list of subject headings reflecting geographic space, time, and thematic keywords of importance to historical researchers.

In this chapter we first examine the importance of maps as tools and sources for genealogists and historical geographers, and the limitations of both paper and digital sources. We explore the potential for GIS as a vehicle enhancing these functions of maps in the historical and genealogical research context. We also provide evidence of overlap and complementarity between sources and needs of these groups of researchers, to support our vision of a GIS-based collaborative tool.[1] This evidence emerges from Dobbs' (2006) mapping and analysis of land grant patterns within a GIS, based on archival document sources describing individual parcels. The first author's (Ruvane) ongoing work investigates the availability and content of other sources discussed below that would more likely be pursued by genealogists and historians than historical geographers.

Though this evidence comes from research focused on central North Carolina's backcountry in the mid eighteenth century, we posit that the basic ideas and the larger vision of a GIS-based collaborative tool would be applicable across the United States as long as suitable adjustments were made regarding regional cadastral systems and the varying timeline of colonization and settlement by Europeans (Price, 1995). For other countries, the basic outline may also be applicable, depending on the availability of archival materials, the state of Internet and computer access, and each country's own cadastral system (Kain, 2002; Kain and Baigent, 1993).

Maps as Sources and Tools

Historical geographers, social historians, and genealogists, as well other scholars concerned with the past, often need to visualize settlement patterns of earlier periods to carry out their research. Yet, for those exploring American history, especially if the time under study is prior to the mid-nineteenth century, few maps exist at

---

1 Though we refer to this vision as "ours" in order to simplify the prose in this co-authored work, the credit for this idea should go to Ruvane (with the enthusiastic support of the second author).

the necessary scale or provide the consolidated picture they seek. To satisfy this information deficiency, many create their own maps, consulting various historical sources to collect geographic clues for reconstructing past neighborhood settings. In the process they may discover new facts that emerge from this synthesis, making maps both source and tool. In this section we discuss briefly the use of maps in settlement reconstruction, the transformation of maps in this digital age, and examine some of the problems that arise for researchers in using both analog and digital maps.

For regions where no suitable maps exist, the ability to illustrate a prior neighborhood setting is invaluable. For example, geographers have long mapped prior communities to explore a range of historical circumstances. Ramsey (1964) reconstructed selected eighteenth century neighborhoods in historic Rowan County, North Carolina, to study the migration motivations and ethnic diversity of settlers who moved from areas in the north to resettle in this southern region. Earle (1975) prepared historic maps to study early economic conditions in All Hallow's Parish of Anne Arundel County Maryland. In a joint project an environmental geographer and biologist reconstructed the location of land settled between 1664 and 1793 in the Gwynns Falls watershed area in Maryland for studying aspects of landscape ecology (Bain and Brush, 2004).

Other scholars who reconstructed previously unmapped historical settings include a local historian and cartographer who rendered several maps to document early landowners who settled in North Carolina (Hughes, 1976a, 1976b, 1977, 1979, 1980). A physicist mapped locations of early settlers' homesteads in North and South Carolina and is a prolific self-publisher of mostly abstracts and indexes to early land records (Pruitt, 1987, 1988). A U.S. Attorney, at the end of his term under the Reagan administration, published a book containing parcel maps and a history of present day Carroll County, Virginia (Alderman, 1985).

Before the late 20th century, maps were typically hand drawn, although they used commercially prepared patterns and photographic techniques on the eve of the digital age. While maps drawn in prior centuries are generally not as dimensionally precise as present-day maps, they often correctly portray the relative position of the features they illustrate (e.g., adjacencies, directional relationships, elevations, etc.). Today, pen and ink have been replaced by computer technology and GIS software in most commercial and scholarly cartography. The impetus for this change was the growing costs associated with maintaining analog (paper) maps, which quickly become obsolete in areas experiencing rapid growth and development. Government and academic institutions have played a key role in this paradigm shift, creating the models and processes that have made possible this new mapping revolution (Mark et al., 1996).

With the advent of affordable computer technology and ongoing technical improvements to GIS applications, the use of electronic maps continues to grow steadily. In the United States much of the growth has been fostered by initiatives from the government's interagency Federal Geographic Data Committee (FGDC). The FGDC was established in 1990 and oversees the National Spatial Data Infrastructure (NSDI) and National Map programs which make public access possible to a broad and diverse set of digital geographic content (Federal Geographic Data Committee, 2005; Office of the President of the United States of America, 2002; USGS, 2005). The NSDI mission is to assure "…that spatial data from multiple sources (federal, state, local, and tribal governments, academia, and the private sector) are available and easily integrated to enhance the understanding of our physical and cultural world." (Office of the President of the United States of America, 2002). Though the NSDI clearinghouse and National Map's interfaces could benefit from improvements in usability, great strides have been made in promoting the best practices employed by these projects.

One far-reaching aspiration of the FGDC is to encourage the production of partner-contributed framework data, representing seven geographic themes recognized by the Committee as fundamental for use in most GIS applications: geodetic control, orthoimagery, elevation, transportation, surface waters, jurisdictional units, and cadastral information (Federal Geographic Data Committee, 2006; Frank et al., 1995; Somers, 1997). The goal of the framework approach is to reduce duplication of efforts across and between user groups by eliminating the need to collect data at various resolutions for the same geographic area. It proposes that each partner concentrate on updating the "...more detailed, accurate, current, and complete data available for an area..." for use as the building blocks of individual framework layers (Somers, 1997). From these detailed layers generalized views (i.e. in smaller scales representing larger areas) could later be generated. Although implementation challenges remain to be resolved, it seems likely this approach will eventually succeed.

Unfortunately for the historical researcher, the primary focus of the FGDC's initiatives, including the framework approach, is maintaining, updating, and sharing data sets representing present-day geographic conditions. Where does this leave those interested in visually exploring local area landscapes of the past through GIS? Several of the geographic themes identified in the FGDC's framework approach are equally important in the study of earlier human activities, but they need to reflect historical conditions and shifting circumstances, ideally by containing historical transportation, surface water, jurisdictional, and cadastral information

Where does one begin if neither analog nor digital maps exist covering one's study area or time period? What about the maps that do exist, but are being overlooked in this GIS revolution, sitting in archives, of little use to researchers working in this new digital world? Maps housed in archives, museums, libraries, or private collections can be hard to find and are often geographically remote. Additionally, the further back in time one goes, the less likely a comprehensive map or one at the appropriate scale even exists. While some U.S. jurisdictions have begun digitizing their older map collections (Mecklenburg County, 2007), most of these are not accessible through the government's NSDI portals, in GIS compatible formats, and few, if any, of these early maps reflect conditions prior to the middle of the nineteenth century. Yet it is evident in academia and cultural heritage institutions that digital historical map data have become increasingly important to their work. Numerous articles point to the untapped potential of incorporating GIS into historical research (Boonstra et al., 2004; Gregory et al., 2003; Knowles, 2002).

Some researchers are independently trying to alleviate the dearth of electronic content and improve accessibility. Several historical atlases have been implemented (Center for Geographic Analysis, n.d.; Fitch and Ruggles, 2003; Hoelscher, 2001; Knowles, 2000; University of Portsmouth, 2007) and a variety of metadata standards have been (Cromley and McGlamery, 2002; West and Hess, 2002) or are being explored (Buckland et al., 2004). Instrumental to interpreting and mapping historical data are gazetteers and thesauri, linking vernacular names to present day terminology and identifying extinct locations and place names (Buckland and Lancaster, 2004; Hill et al., 2002). Distribution of historical GIS information is also being addressed (UC Santa Barbara, 2007; University of California, 2004; Lim et al., 2002; Rumsey, 2001, 2003; Zerneke, 2003), although many efforts presently represent isolated initiatives that are limited in scope and lack a centralized distribution network for combining or searching across similar endeavors.

Clearly progress is being made, yet the focus of most historical GIS initiatives is on generalized map data, suited for small scale representations used for portraying aggregated statistics at the state or national level. These generalized spatial datasets fall short of the detail necessary for visualizing neighborhood dynamics, and because larger scale representations typically do not exist, the only recourse for an historical researcher is to find and consolidate evidence suitable for reconstructing the landscapes they wish to study. Especially when the interest lies in establishing the location of backcountry neighborhoods formed prior to the mid-nineteenth century; the imprecise content contained in early American source material, lack of suitable reference maps, and absence of large-scale property maps leaves many gaps.

Critical to the process of reconstructing historical settlement patterns under these conditions is the collection, organization, and interpretation of information to guide in the positioning of land parcels as the researcher creates his or her own map. Common evidence sought by historical geographers and genealogists to satisfy this information need include documenting the names of settlers and their neighbors, size and description of properties, and proximity of each parcel to nearby geographic and cultural features, both historical and contemporary. Perhaps the most valuable evidence for mid-nineteenth century or earlier research is contained in original land grant records, which may include much of the evidence sought. Supplemental geographic information can be found in archived material such as legal documents, organizational reports, and privately held family records.

The use of land records for reducing map-related information gaps is not uncommon; such records have been used to inform a variety of historical GIS research projects. For example, Bain and Brush (2004) presented a technique for reconstructing original land ownership patterns based on colonial era warrant and patent survey records. Mires (1993) employed a GIS to reconstruct Louisiana's colonial land claims to evaluate settlement patterns in relation to the potential natural vegetation. Healey and Stamp (2000) identified theoretical, methodological and practical considerations specific to regional scale GIS applications, highlighting the value of land record evidence and technical issues.

However, for neighborhood or regional scale investigation of past human spatial activity to be successful, a variety of historical framework datasets are needed as background to historical cadastral data. Fortunately some jurisdictional layers depicting the frequent shifts in U.S. county boundaries are available in a series produced under the guidance of John Long (About the Atlas, 2006) and supported by the Newberry Library. Although the initial printed volumes did not employ GIS production techniques, historical GIS boundary data sets are available for 15 states as of this writing with others expected to follow (Newberry Library, 2007). An ongoing project led by the Minnesota Population Center is working on reconstructing digital census tract boundaries, but at this time only 1990 geographic units appear available for selected states (*National Historic GIS*). From our earlier discussion of the FGDC's fundamental background layers three themes critical to historical research—transportation, watercourses, and cadastral layers—remain in need of digital historical reconstruction. Although individual researchers often do this for the limited areas and themes of interest to them, most were not designed with the foresight to enable sharing.

For those without the ability to carry out their own digital historical reconstructions in GIS, it is clear that the usefulness of this technology is limited in the absence of a larger public-access framework. We turn now to a discussion of existing public GIS frameworks and an elaboration of our vision for a historical collaborative tool for genealogists, historical geographers, and other researchers of the past.

GIS as a Cross-Disciplinary Research Tool

While no large-scale collaborative GIS tool currently exists, related studies and ongoing initiatives point to viable aspects for consideration in the development of such an historical GIS. Museums have been exploring tools for implementing two-way engagement with their public for improving access to their collections (Bearman and Trant, 2005); others have explored methods for incorporating GIS tools into digital libraries (Boxall, 2003; Chen et al., 2005; Crane and Wulfman, 2003; Smith and Crane, 2001); an historical geographer highlights contributions his field could offer (Holdsworth, 2003). Tools to facilitate complex and long-term geospatial collaboration already exist (MacEachren et al., 2005; Schafer et al., 2005) and the value of offering online access to archival material is recognized (Cruikshank et al., 2005; Rumsey, 2001; Tibbo, 2003). Improvements in locating historical material that provide links to related maps are also being implemented (Buckland et al., 2004; Feinberg, 2003).

What is missing is a holistic approach to the historical researchers' investigation process. Tools are needed not only for finding suitable historical content, both maps and text, but for collecting and organizing evidence, creating maps based on the findings, and finally as a method to publish and share one's results. Ideally such a tool could accommodate a diverse population, from professionals involved with historical research to lay people interested in visualizing early American neighborhoods Foundational to this diverse group would be a geographic interface providing access to regional historical cadastral maps, along with links to primary source material and research notes that authenticate and explain the images presented.

A contributory interactive research and cadastral mapping tool would offer not only efficiencies in locating and sharing historical materials, but the ability to challenge or expand upon past findings. For example, a family historian for a region in North Carolina stated his map of original land grants represented the earliest grants to settlers in the area, yet further investigation using Dobbs' (2006) research for comparison, showed that this was not the case. In reality his map illustrated the location of the first land grants issued by the state; earlier grants had been issued by the previous Colonial administration. Despite this error in documentation, the maps were relatively accurate for the content they portrayed and once placed in context could serve to inform others. The value to archives and libraries, and in turn future researchers, would be the expanding links to related material discovered by researchers during the course of their investigations. Additionally, new keyword contributions would considerably enhance most institution's sparsely indexed historical material and form the basis for constructing invaluable thesauri and gazetteers.

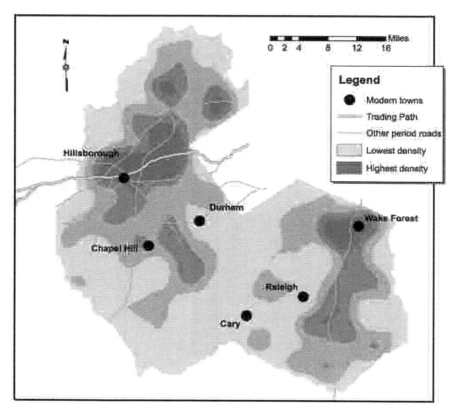

Kernel density reclassified (with period roads and modern towns)

Figure 3.2 GIS output in the form of spatial pattern analysis of land parcels (Dobbs, 2006)

GIS is of course used by many historical geographers for analysis of spatial patterns, moving beyond mapping. Yet often the initial stages of such GIS work require mapping of land parcels or other entities derived from non-map sources such as land records. In Dobbs' (2006) research, for example, though the final GIS output was in the form of spatial and temporal pattern analysis maps (see Figure 3.2), these

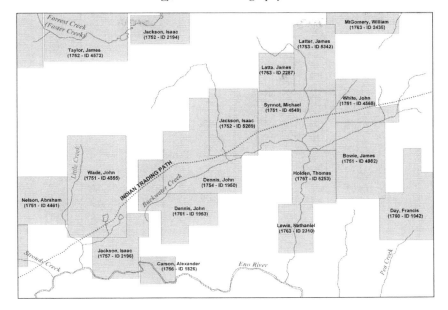

Figure 3.3 Large-scale GIS land parcel mapping based on archival sources

depended on the mapping of individual land parcels, which in turn depended on the reconstruction of historical county boundaries (see Figure 3.3). These aspects of the GIS framework she created in the research process represent the kind of foundation required for implementing our vision. This is especially true as the land parcel information was initially gathered from the archival sources into a database designed for this purpose by the first author. It is this marriage of spatial information and attribute information which makes GIS so powerful, and which is essential to the interactive nature of the collaborative tool we envision. In the next section, we present evidence that data sources likely to be used by genealogists, local historians, and some historical geographers overlap and complement each other enough to make such a tool viable and useful to both groups as well as others.

Evidence for Overlap and Complementarity

This chapter draws on a larger research endeavor aimed at assessing the information needs, viability, and potential level of interest for the collaborative national mapping tool we described. The larger project explores the information needs of historical geographers and genealogists, specifically information of value for geographically positioning tracts of land occupied in central North Carolina during the mid-eighteenth century. In this chapter we compare the evidence culled from land grant documents by Dobbs (2006) in her historical geographical research to clues contained in alternate source materials representative of genealogical or local history accounts covering the same time period and region. The aim was to evaluate the basic building blocks of these respective research approaches: the spatially related information they seek, or choose to document, and the sources consulted for piecing together the mosaic of prior neighborhoods.

The objective of what we present here was to answer four questions:

- *What research overlaps exist between historical geographers and genealogists?* (i.e., how similar are their information needs and objectives?)
- *Is there a definitive stakeholder for a particular type of evidence?* (i.e., does one group focus on capturing certain categories of evidence unique to their research effort, such as geographic locations, historical context, or interpersonal relationships? Which group is more likely to accurately maintain and expand upon a particular type of evidence collected?)
- *Is there complementary evidence collected by one group that could facilitate the process of the others' map reconstruction or historical investigation?*
- *Do the study's findings indicate whether a shared tool could be successful?* (i.e., would substantial duplication of effort be eliminated if an authoritative "annotation tool" were available, allowing each to concentrate their efforts on contributing the data most important to them?)

Dobbs' (2006) research was aimed primarily at teasing out the influence of the Indian Trading Path, a road of indigenous origin, on colonial settlement patterns in the Piedmont of North Carolina. Her approach to this involved examining the earliest documented European land claims in the area, which in turn required engaging with the thousands of individual land records extant for the period and area and constructing a map from these data. The primary sources of most interest to her in this process were the land records themselves, dating from 1748 to 1763, and eighteenth century maps of North Carolina.

In working with colonial-era land records in North Carolina, a researcher may encounter up to four (or five if there is assignee paperwork) individual documents related to the acquisition of a particular property: an entry, a warrant, a survey, and a grant of deed or patent. These records provide various clues for reconstructing the geographic position of a tract of land. At minimum they include transaction dates and the names of the grantor and grantee(s), but the most helpful documents contain the names of neighbors and locations of proximate geographic and cultural features. Especially critical are the survey documents which contain a metes and bounds narrative and small parcel map (e.g. plat) illustrating the property's boundaries (see Figure 3.4). Survey documents were filed both as records in their own right, and often again as part of the final grant of deed, but not always. In North Carolina, these records are held by the NC State Archives, accessible to the public in microfilm format. They are (unfortunately for the geographer) organized by name, within the county in which the land was situated at the time, adding to the difficulty of using them for this type of research.

Dobbs (2006) chose to work with all four (or five) of the land grant record types, extracting clues from them to clarify the associations between individual documents describing the same tract of land. Unlike researchers before her who tended to work solely with deed records, she elected this more thorough approach because it

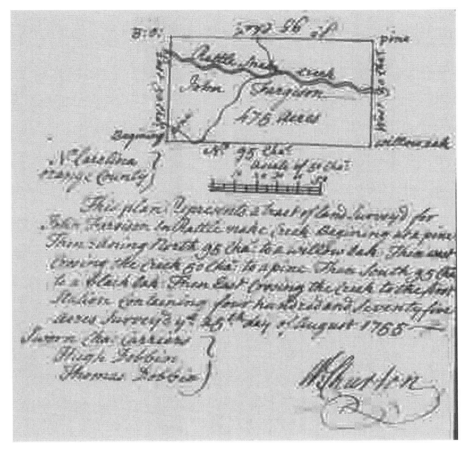

Figure 3.4 A typical 18th century survey from North Carolina's Granville District

was clear not all occupied land had a consistent paperwork trail. Some records in the sequence are missing or damaged; claims at times were abandoned midstream through the process; and in other cases a dispute may have arisen with the outcome recorded elsewhere, leaving a gap in the chain of acquisition. This turned out to be a good decision, because the individual documents related to a particular tract of land frequently contained complementary information not included in prior or later transactions, together providing the information needed to properly position a tract of land.

Many other potential sources had to be ignored because of the length of time required to deal properly with the land records, and this is where Dobbs (2006) could have benefited from the complementary knowledge of genealogists in a collaborative, interactive, online tool. Settlement reconstruction can be enhanced by using not just the evidence in the land documents, but information from other kinds of documents such as wills, secondary deed transfers, and family papers. Many such documents are more people-oriented than land-oriented, and so may initially be of less use to an historical geographer than a genealogist, but the complementary evidence they contain could prove instrumental in resolving an uncertainty related to the place an individual lived, ultimately helping in the mapping process.

Genealogists deal with perhaps the broadest and most diverse assortment of historical evidence in order to reconstruct their family's lineage. Unlike historians and historical geographers, their investigations are rarely restricted to one time period or location, but require broad knowledge of the types of source material available for a specific locality during a particular century. Many travel wherever the clues lead them, tracing the migration routes of their ancestors and searching for proof of their family's heritage. Genealogists have been identified as one of the largest group of patrons frequenting archives, government record offices, and historical societies in search of historical evidence (Ailes and Watt, 1999; Boyns, 1999; Forbes and Dunhill, 1996).

The first author has been examining a range of materials relevant to Dobbs' research period and area, but not exploited in that research, focusing on family collections held by the Southern Historical Collection at the University of North Carolina rather than the official records held by the NC State Archives, and on secondary materials including historical accounts, indexes, compiled genealogies, and compiled land grant maps.[2] Family collections can be especially important because they may contain a variety of useful documents from the time period under study, including estate or neighborhood maps, yet these types of important materials are often not indexed within library catalogs or archival finding aids. From 43 such manuscript and secondary sources examined, Ruvane's current unpublished work draws from three to present a content analysis (Figures 3.5 and 3.6) designed to identify overlap and complementarity of content between representative genealogical sources and the land records used by Dobbs (2006). These three were chosen because they had a clear geographic connection to each other, covering a small area within the southwest region of Dobbs' study area, and because they were indicative of material found in special archives. One was a typed genealogy of a prominent local family's lineage including the names, dates, and locations of major events in their lives (Alexander Family, 1952); the other two were historical monographs written by respected local historians (Alexander, 1897; Sommerville, 1939).

For this comparison a relational analysis technique, also called semantic analysis, was employed to explore term relationships that link people to the locations they inhabited, based on methods commonly employed in content-analysis (Palmquist et al., 1997). Two "term maps" were developed, each drawn from a different set of textual material. Figure 3.5 explores the content contained in the eighteenth century land grant records analyzed by Dobbs, the central relationship based on units of land. Figure 3.6 draws from the genealogical and historical account material described above, the pivotal relationship based on people. In both term maps the goal was to look for overlapping and complementary information that might help to inform the reconstruction of an historical map of settlement patterns.

---

2   The authors are indebted to Maria Asencio who spent hours verifying suitable content while participating in The University of North Carolina at Chapel Hill, School of Education, 2005 Summer Pre-Graduate Research Experience Program (SPGRE).

At the time of preparing this analysis the total number of land grant documents Dobbs had entered into her database totaled 3,960. Multiple queries were performed to extract and quantify the underlying relationship terms associated with these land records, shown in Figure 3.5. The numbers in parentheses following each term associated with a relationship type indicate the unique occurrence of that term within an individual document. For example, if two or more distinct water features were recorded in one document, the term *water* was only counted once. An exception to this counting method was used for the category *"is associated with historical county"*, if a parcel had resided in more than one *historical county* over the course of time each unique county name was counted.

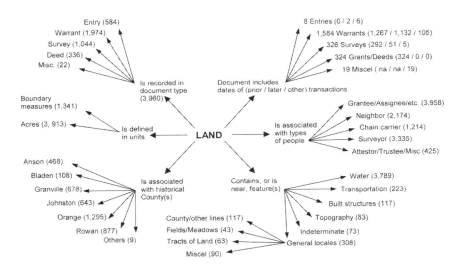

Figure 3.5   18th century land records links—LAND to key relationships

For the genealogical term map, shown in Figure 3.6, the numbers in parentheses following the terms associated with a relationship type indicate the occurrence of each distinct term as they relate to an individual person. For example, if a person lived near two or more historical or later built structures, or had multiple offspring, only one instance of this type of relationship was counted. As in the land term map, an exception was taken in the count of unique county relationships (e.g., *resided within historical and/or later county/defined neighborhoods*), each unique county name or neighborhood a person was associated with was counted.

As the figures indicate, there are clearly information synergies between the two types of records analyzed. While the primary message of land grant records concerns the characteristics of a tract of land, and the genealogical or historical accounts selected focus on people, they both contain overlapping evidence. Common to both are the names of people, geographic or cultural features, counties, and date evidence essential for validating comparable time lines. The differences occur in the level of detail and the perspective taken by each researcher.

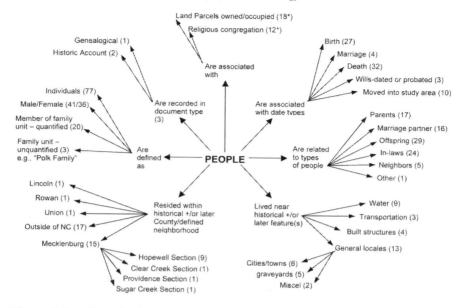

Figure 3.6   Genealogical/historical material links—PEOPLE to key relationships

The sharing of evidence contained in land grant records and genealogical or historical accounts could improve the process of reconstructing a past neighborhood setting. The surveys, or deed records with surveys, contain the critical measurements needed for reconstructing the shape and size of individual parcels. This important evidence was absent from the genealogical material reviewed. What the land records lacked was evidence important for identifying a person's uniqueness (e.g., birth, marriage, death, date of migration, community affiliations, etc.), which at times could prove instrumental in the mapping process. For example, not all land records identify historical features, such as a nearby watercourse or landmark, needed for positioning a parcel into its prior location. Genealogical material may provide this type of information, or contain supplemental evidence such as a person's church affiliation that could help to pinpoint a parcel's location.

An important benefit of genealogical or historical accounts is the integration of contemporary knowledge with historical facts, linking geographical and cultural features to present day circumstances. Knowledge of an historic parcel's location in relation to its present day county, or proximity to contemporary cultural features, would help to narrow the administrative area in which a tract might have resided. However, many of the genealogical accounts failed to identify whether they were speaking in terms of historical or contemporary settings which at first can be confusing. For example one early source (Alexander, 1897) defined the location of an eighteenth century homestead in terms of its proximity to nineteenth century neighbors, who today no longer exist, and most of the people identified as

Mecklenburg County residents resided in Anson County during the period under discussion. Nevertheless the complementarity of these sources and the land records used by Dobbs was confirmed.

The content analysis maps indicate definitive stakeholders in primarily two categories. Unique to Dobbs's information needs were accurate parcel boundary measurements to facilitate reconstruction of each individual tract of land for placement on a map. This information was absent in the genealogical or historical accounts analyzed. In the genealogical and historical records, dates related to an individual such as their birth, death, or marriage along with familial relationships appeared to be of primary importance. Both groups collected data concerning neighbors, cultural and geographic features, and administrative boundaries, although genealogical records offer correlations to contemporary locations not available in land records.

Conclusion

The complementary evidence contained in these different record types could prove invaluable in forwarding each other's respective historical investigation. An historical geographer may have positioned a tract of land associated with an indistinct person yet be missing information pointing to likely neighbors needed for completing the puzzle. A genealogist may have supporting evidence containing just the clues needed, such as the unique identity of a landholder and information on his or her community relationships that could fill in the missing pieces. Sharing such information would forward not only the historical geographer's mapping objective, but also the genealogist's inquiry into the neighborhood dynamics of areas settled by his or her ancestors.

A shared tool seems promising based on these findings. There appears to be substantial duplication of effort that could be eliminated, if historical geographers and genealogists had a suitable vehicle through which to share their overlapping as well as complementary evidence. Links providing authentication of a person's unique identity could help to substantiate a historical geographers research, and to resolve ambiguous locations often found in original land grant documents. For the genealogist, access to authoritative maps prepared by historical geographers could aid in their hunt for additional primary source material, leading to the discovery of new relations along the family tree. Furthermore, as relationships among documents and places were solidified, and more documents images could be linked to places within the GIS interface, more researchers would be able to utilize the material and to add their own insights or additional linkages. In this way, the overall effectiveness of the tool would continue to increase over time.

References

About the Atlas of Historical County Boundaries (2006), <http://www.newberry.org/ahcbp/atlasabout.html> Accessed 1 May 2007.

Ailes, A. and Watt, I. (1999), "Survey of Visitors to British Archives, June 1998", *Journal of the Society of Archivists*, 20(2): 177-194.

Alderman, J.P. (1985), *Carroll 1765-1815, the Settlements: A History of the First Fifty Years of Carroll County, Virginia*. Hillsville, VA: Alderman Books.

Alexander, J.B. (1897), *Biographical Sketches of the Early Settlers of The Hopewell Section and Reminiscences of the Pioneers and their Descendants by Families*, Available in the North Carolina Collection, Wilson Library, University of North Carolina, Chapel Hill.

Alexander Family (1952), *Alexander Family Genealogy #2853 (1952)*, Manuscript available in the Southern Historical Collection, Wilson Library, University of North Carolina, Chapel Hill.

Bain, D.J. and Brush, G.S. (2004), "Placing the Pieces: Reconstructing the Original Property Mosaic in a Warrant and Patent Watershed", *Landscape Ecology*, 19(8): 843-856.

Baugh, I.W. (2004), "Baugh Families of Logan County, 1817-1880", *Kentucky Ancestors*, 40(1): 26-27, 30-32.

Bearman, D. and Trant, J. (2005), "Social Terminology Enhancement Through Vernacular Engagement: Exploring Collaborative Annotation to Encourage Interaction with Museum Collections", *D-Lib Magazine*, 11(9): n.p.

Boyns, R. (1999), "Archivists and Family Historians: Local Authority Record Repositories and the Family History User Group", *Journal of the Society of Archivists*, 20(1):62-72.

Boonstra, O., Breure, L., and Doorn, P. (2004), *Past, Present and Future of Historical Information Science*. Amsterdam: Netherlands Institute for Scientific Information.

Boxall, J. (2003), "Geolibraries: Geographers, Librarians and Spatial Collaboration", *The Canadian Geographer*, 47(1): 18-27.

Buckland, M.K., Gey, F.C. and Larson, R.R. (2004), *Going Places in the Catalog: Improved Geographic Access [Final Report]*. Berkeley, CA: Electronic Cultural Atlas Initiative, University of California.

Buckland, M.K. and Lancaster, L. (2004), "Combining Place, Time, and Topic: The Electronic Cultural Atlas Initiative", *D-Lib Magazine*, 10(5): n.p.

Center for Geographic Analysis (n.d.), "China Historical GIS", Available http://www.fas.harvard.edu/~chgis/ (accessed 27 April 2007).

Chen, C.C., Wactlar, H.D., Wang, J.Z. and Kiernan, K. (2005), "Digital Imagery for Significant Cultural and Historical Materials: An Emerging Research Field Bridging People, Culture, and Technologies", *International Journal on Digital Libraries*, 5: 275-286.

Crane, G. and Wulfman, C. (2003), *"Towards a Cultural Heritage Digital Library"*, Paper presented at the International Conference on Digital Libraries, Houston, TX.

Cromley, R.G. and McGlamery, P. (2002), "Integrating Spatial Metadata and Data Dissemination Over the Internet", *International Association for Social Science Information Service and Technology Quarterly*, 26(1): 13-16.

Cruikshank, K., Daniels, C., Meissner, D., Nelson, N.L. and Shelstad, M. (2005), "How do We Show You What We've Got? Access to Archival Collections in the Digital Age", *Journal of the Association for History and Computing*, 3(2). Online, retrieved from http://mcel.pacificu.edu/JAHC/JAHCVIII2/articles/cruikshank.htm#12.

Dobbs, G.R. (2006), *The Indian Trading Path and Colonial Settlement Development in the North Carolina Piedmont,* Unpublished PhD dissertation, University of North Carolina, Chapel Hill.

Earle, C.V. (1975), *The Evolution of a Tidewater Settlement System: All Hallow's Parish, Maryland, 1650-1783.* Unpublished Research paper, University of Chicago, Chicago, IL.

Federal Geographic Data Committee (2005), "*Executive Order 12906: The National Spatial Data Infrastructure*", Available http://www.fgdc.gov/nsdi/policyandplanning/executive_order (accessed 28 February 2007).

Federal Geographic Data Committee (2006), "Framework Overview", Available http://www.fgdc.gov/framework/frameworkoverview (accessed 27 April 2007).

Feinberg, M. (2003), *Application of Geographical Gazetteer Standards to Named Time Periods*, draft report, Berkeley, CA: Institute of Museum and Library Services.

Fitch, C.A. and Ruggles, S. (2003), "Building the National Historical Geographic Information System", *Historical Methods*, 36(1): 41-51.

Forbes, H. and Dunhill, R. (1996), "Survey of Local Authority Archive Services: 1996 Update", *Journal of the Society of Archivists*, 18(1): 37-57.

Frank, S.M., Goodchild, M.F., Onsrud, H.J. and Pinto, J.K. (1995), *Framework Data Sets for the NSDI* (NCGIA Technical Papers Series, Technical Report 95-1). Santa Barbara, CA: National Center for Geographic Information and Analysis.

Greenwood, V.D. (2000), *The Researcher's Guide to American Genealogy* (3rd edn.), Baltimore, MD: Genealogical Publishing Co.

Gregory, I.N., Kemp, K. and Mostern, R. (2003), "Geographical Information and Historical Research: Current Progress and Future Directions", *History and Computing*, 13: 7-22.

Healey, R.G. and Stamp, T.R. (2000), "Historical GIS as a Foundation for the Analysis of Regional Economic Growth: Theoretical, Methodological, and Practical Issues", *Social Science History*, 24(3): 575-612.

Helmbold, F.W. (1976), *Tracing Your Ancestry: A Step-by-Step Guide to Researching Your Family History*, Birmingham, AL: Oxmoor House.

Hill, L.L., Hodge, G.M. and Smith, D.A. (2002), "*Workshop Agenda: Digital Gazetteers: Integration Into Distributed Digital Library Services*", Paper presented at the JCDL 2002 NKOS Workshop, Portland, Oregon, 18 July.

Hill, M.E.V. (2006), "*DeedMapper: Reconstruct Your Ancestor's Land*", Paper presented at the Utah Valley Personal Ancestral File Users Group: Monthly Meeting Presentation, Provo, Utah, 10 June.

Hoelscher, S. (2001), "Mapping the Past: Historical Atlases and the Mingling of History and Geography", *The Public Historian*, 23(1): 75-85.

Holdsworth, D.W. (2003), "Historical Geography: New Ways of Imaging and Seeing the Past", *Progress in Human Geography*, 27(4): 486-493.

Hughes, F. (1976a), *Montgomery County, North Carolina: Historical Documentation*, Jamestown, NC: Custom House.

Hughes, F. (1976b), *Randolph County, North Carolina: Historical Documentation No. 2*, Jamestown, NC: Custom House.

Hughes, F. (1977), *Davie (the forks of the Yadkin) County, North Carolina: Historical Documentation No. 4*, Jamestown, NC: Custom House.

Hughes, F. (1979), *Rockingham County, North Carolina: Historical Documentation No. 7*, Jamestown, NC: Custom House.

Hughes, F. (1980), *Guilford County, North Carolina: Historical Documentation No. 9*, Jamestown, NC: Custom House.

Kain, R.J.P. (2002), "The Role of Cadastral Surveys and Maps in Land Settlement from England", *Landscape Research*, 27(1): 11-24.

Kain, R.J.P. and Baigent, E. (1993), *The Cadastral Map in the Service of the State: A History of Property Mapping*, Chicago: University of Chicago Press.

Kashuba, M. (2005), Charting a Research Course Using Maps and Geography. *NGS News Magazine*, 31(4): 26-30.

Knowles, A.K. (2000), "Mapping Wisconsin", *The Geographical Review*, 90(2): 277-284.

Knowles, A.K. (2002), *Past Time, Past Place: GIS for History*, Redlands, CA: ESRI Press.

Lim, E.P., Goh, D.H.L., Liu, Z., Ng, W.K., Khoo, C.S.G. and Higgins, S.E. (2002), "G-Portal: A Map-Based Digital Library for Distributed Geospatial and Georeferenced Resources", Paper presented at the International Conference on Digital Libraries, Portland, Oregon.

MacEachren, A.M., Cai, G., Sharma, R., Rauschert, I., Brewer, I., Bolelli, L., Shaparenko, B., Fuhrmann, S. and Wang, H. (2005), "Enabling Collaborative Geoinformation Access and Decision-Making Through a Natural, Multimodal Interface", *International Journal of Geographical Information Science*, 19(3): 293-317.

Maptech (2007) *Historic U.S. Geological Survey Maps*. Available http://historical.maptech.com/ (accessed 28 February 2007).

Mark, D.M., Chrisman, N., Frank, A.U., McHaffie, P.H. and Pickles, J. (1996), *The GIS History Project*. Unpublished report, available http://www.ncgia.buffalo.edu/gishist/bar_harbor.html (accessed 28 February 2007).

Mecklenburg County (2007), "Mecklenburg County Historical Map Archive", Available http://www.charmeck.org/Departments/Geospatial+Information+Services/Historical+Maps+Library/Home.htm (accessed 02 April 2007).

Mires, P.B. (1993), "Relationships of Louisiana Colonial Land Claims with Potential Natural Vegetation and Historic Standing Structures: A GIS Approach", *The Professional Geographer*, 45(3): 342-350.

Newberry Library (2007), "Atlas of Historical County Boundaries: Downloadable Files", Available http://www.newberry.org/ahcbp/download_index.html (accessed 30 November 2007).

Office of the President of the United States of America (2002), "*Circular No. A-16 Revised*", Available http://www.whitehouse.gov/omb/circulars/a016/a016_rev.html#4 (accessed 28 February 2007).

Palmquist, M.E., Carley, K.M. and Dale, T.A. (1997), "Applications of Computer-Aided Text Analysis: Analyzing Literary and Nonliterary Texts", in C.W. Roberts (ed.) *Text Analysis for the Social Sciences: Methods for Drawing Statistical Inferences from Texts and Transcripts*, Mahwah, NJ: Lawrence Erlbaum, pp. 171-189.

Price, E.T. (1995), *Dividing the Land: Early American Beginnings of our Private Property Mosaic*, Chicago: University of Chicago Press.

Pruitt, A.B. (1987), *Index to South Carolina County Maps: Anderson 1897, Greenville 1882 & 1904, Greenwood 1898 & Spartanburg 1887*, Spartanburg, SC: A.B. Pruitt.

Pruitt, A.B. (1988), *Index to North Carolina County Maps: Davie, Guilford, Montgomery, Randolph, Rockingham, Stokes, Surry, & Yadkin*, Guilford, NC: A.B. Pruitt.

Ramsey, R.W. (1964), *Carolina Cradle: Settlement of the Northwest Carolina Frontier, 1747-1762*, Chapel Hill, NC: University of North Carolina Press.

Rumsey, D. (2001), "*Historical Map Collection Web Site*", Paper presented at the Museums at the Web 2001 Conference, Seattle, Washington, 14-15 March.

Rumsey, D. (2003), "*David Rumsey Map Collection*", Available http://www.davidrumsey.com/ (accessed 28 February 2007).

Schafer, W.A., Ganoe, C.H., Xiao, L., Coch, G. and Carroll, J.M. (2005), "Designing the Next Generation of Distributed, Geocollaborative Tools", *Cartography and Geographic Information Science*, 32(2): 81-100.

Smith, D.A. and Crane, G. (2001), "*Disambiguating Geographic Names in a Historical Digital Library*", Paper presented at the Research and Advanced Technology for Digital Libraries: 5th European Conference, Darmstadt, Germany.

Somers, R. (1997), *Framework Introduction and Guide*, Reston, VA: Federal Geographic Data Committee.

Sommerville, C.W. (1939), *History of Hopewell Presbyterian Church for 175 Years From the Assigned Date of its Organization, 1762*, Charlotte, NC: Hopewell Presbyterian Church.

Tibbo, H.R. (2003), "Primarily History in America: How U.S. Historians Search for Primary Materials at the Dawn of the Digital Age", *American Archivist*, 66(1): 9-50.

UC Santa Barbara (2007), "*Center for Spatially Integrated Social Science: Spatial Resources for the Social Sciences*", Available http://www.csiss.org/ (accessed 28 February 2007).

University of California (2004), "*Alexandria Digital Library*", Retrieved http://www.alexandria.ucsb.edu/ (accessed 28 February 2007).

University of Portsmouth (2007), "Great Britain Historical Geographical Information System (GBHGI)", Available http://www.port.ac.uk/research/gbhgis/ (accessed 27 April 2007).

USGS (2005), "*The National Map: The Nation's Topographic Map for the 21st Century*", Available http://nationalmap.gov/nmabout.html (accessed 28 February 2007).

Using Maps in Genealogy (2006) <http://erg.usgs.gov/isb/pub/factsheets/fs09902.html> Accessed 30 April 2007.

West, L.A. and Hess, T.J. (2002), "Metadata as a Knowledge Management Tool: Supporting Intelligent Agent and End User Access to Spatial Data", *Decision Support Systems*, 32(3): 247-264.

Zerneke, J. (2003), "*Electronic Cultural Atlas Initiative*", Available http://www.ecai.org/ (accessed 01 October 2003).

Chapter 4

A Genealogy of Environmental Impact Assessment

William Hunter

Introduction

On 4 October 2006, representatives of the United States Federal Highway Administration and the Pennsylvania Department of Transportation met with local historians and genealogists at a small historical society library in Berlin, Pennsylvania. The purpose of the meeting was to discuss strategies for minimizing and mitigating the adverse effects of a proposed major new four-lane highway on a rural historic landscape. The meeting being held at the genealogical building of the Berlin Area Historical Society was appropriate given the degree to which genealogical knowledge shaped the outcome of the project development process, affecting not only the location of the proposed highway, but also its form and appearance.

The work of genealogists, often marginalized within the academy, has become increasingly important to the practice of cultural resource management (CRM) and the identification and evaluation of historic properties during the development of Federal undertakings. Cultural resources, including broad landscapes, are often a major constraint to the development of federally funded projects such as highways. This chapter explores how CRM involves the negotiation of complex landscapes and place-identities that genealogists create, reinforce and disseminate through local historical societies, publications, websites and in public forums. This survey concludes that genealogy not only grounds identity, prompts travel and tourism, and creates a sense of belonging, but can also affect the form and location of the most material of landscapes, a modern interstate highway (Nash, 2000b; Meethan, 2004).

The U.S. 219 Improvements/S.R. 6219, Section 020 highway project is located within a region historically shaped by the settlement of German, Pennsylvania German, and Scots-Irish farmers, and later reshaped through the large-scale extraction of coal and out-migration. The project is located within an area once well known as the Brothers Valley or *Bruedersthal*, a spatially discrete culture region. The project entails the improvement of a 15.2-mile section of the two-lane highway, and ultimately involves the construction of a limited access highway on a new alignment between the boroughs of Somerset and Meyersdale in Summit, Brothers Valley, and Somerset Townships, Somerset County. The stated purpose of the project is to improve safety, access, and traffic conditions, while supporting regional economic development initiatives.

The project is located in an area known for the rich quality of the cultural landscape and as an area steeped in historic tradition, a mosaic reflecting the overlapping phases of Somerset County's historical development (Glessner, 2000: 7; Summers et al., 1994). The collage reflects distinct eras in the history and development: the construction of the pre-settlement landscape, the agricultural landscapes of the settlement eras, the nationalizing landscape of the railroad era, and the automobile era's landscape of mobility. Further, this layered landscape, affected by generations of coal mining, has entered the popular imagination as the crash site of United Airlines Flight 93 on September 11, 2001.

History and Highways

The National Historic Preservation Act, the National Environmental Policy Act, and portions of the Federal Transportation Act share the premise that the preservation of properties significant to the nation's heritage is in the public interest. These fundamentally progressive laws require Federal agencies involved in an undertaking to identify, evaluate and avoid historic properties and consider the views of the public, a broadly defined category that includes historical societies, advocacy organizations and individuals. What is, and what is not, considered historically significant is often grounded in the work of local historians and genealogists and their capacity to deploy a reading of the past, often a nostalgia, in the service of a particular interest or position (Lowenthal, 1977). To meet the requirements of the National Historic Preservation Act, Federal agencies employ professional historians and archaeologists to identify historic properties in a practice widely known as CRM.

The three laws, though distinct, govern the development of major transportation projects. Section 106 requires Federal agencies to take into account the effect of their undertakings on any district, site, building, structure or object that is included in or eligible for inclusion in the National Register of Historic Places. In addition, Section 4(f) of the Department of Transportation Act of 1966 requires that federally funded transportation projects avoid the use of land from historic properties if there is no feasible and prudent alternative to the use of that land, and that agencies use all possible efforts in planning to avoid or minimize project effects on historic properties. Historic properties may factor in an analysis of project impacts and the selection of a preferred alternative during the National Environmental Policy Act (NEPA) process. The implementation of these laws has an enormous influence on the high stakes development of federally funded or permitted projects.

The National Historic Preservation Act affords a measure of protection for properties that are listed on or are eligible for inclusion on the National Register of Historic Places. Under the terms of the relevant regulation (36 CFR 60), properties are eligible for inclusion if they are associated with significant events or patterns of events; are associated with individuals considered significant to our past; are important examples of a building type, period or method of construction; represent significant movements within American architecture; or are the work of a master; and/or have yielded, or may yield, information important in history or prehistory (National Park Service, 1996). To be eligible, significant properties must also retain integrity, which refers to the ability of the property to convey, through distinctive physical characteristics, the qualities that constitute its historical significance. Only properties that are historically significant *and* retain integrity are eligible for the National Register of Historic Places.

The significance of historic properties is rooted in the historic context, recognition that resources, properties, or happenings in history "do not happen in a vacuum but are part of larger trends or patterns" (National Park Service, 1996: 7; Secretary of the Interior, 1983: 9). A well-defined context is critical to the accurate evaluation of significance which is critical to assessing effects on historic properties. Investigators define historic contexts from "existing information, concepts, theories, models, and descriptions," often relying heavily on work of local historians and genealogists and the views of the public (Secretary of the Interior, 1983: 8). Involvement of the public has broadened the definition of historic properties to include traditional cultural properties, something of a collective genealogy of a community in place (Anderson, 1998).

Cultural Resource Management

The laws and regulations that mandate and structure CRM are social regulatory mechanisms that guide the meaningful re-creation of cultural landscapes (Peet, 1996). Like the broader heritage movement, CRM is also a site where the production of symbols is intimately connected to both the production of jobs and the production of space (Zukin, 1997). If power entails the creation and manipulation of social signification, the making of meaning, then CRM is clearly an avenue of power (Osborne, 1998; Soja, 1997). By recreating landscapes, filling them with signs, preserving some artifacts and destroying others, CRM practitioners produce meaning on the landscape and thereby affect the everyday behavior of the people who call these subtly manufactured places their home (Harvey, 1979; Peet, 1996: 23).

The practice of CRM often involves negotiation of the landscapes that link material places and the imagination. In fact, the very processes of environmental impact assessment negotiate a collage of geography, memory, and sentiment informed by the work of local historians and genealogists (Tuan, 1977: 729). The environmental process in fact is broad enough to include imaginative historical geographies, the geographies of selective memory and genealogies that collude to form the sense of place. The production and dissemination of place-bound genealogies creates iconic elements in the landscape that frame the geographies of everyday life and anchor historical memory, demonstrating the power of genealogical knowledge when tied to a specific place or artifact (Holdsworth, 1997; Johnson, 2004).

CRM reflects, interprets and produces public memory, geographical knowledge and commemorative landscapes within a public process (Bodnar, 1992; Osborne, 1998). The embrace of public memory as a justification of historical significance can be undermined by constructed authenticity, invented traditions, and commercialized heritage culture (Harvey, 1993: 12; Hobsbawm, 1983). Because of this complexity, CRM must critically approach both state-sanctioned forms of public memory and popular forms of memory produced through genealogical imaginations (Nash and Graham, 2000).

CRM is in fact a process of negotiation of the symbolic and material worlds that would benefit from an infusion of informed geographic perspectives (Nash and Graham, 2000: 5). Dominated by disciplines such as architectural history and archaeology, geographers are under-represented in this fundamentally spatial practice of CRM (King, 1998). As a result, CRM surveys, under-informed by geography's insights into the production of landscape or its sensitivity to the problematic notion of place, often overlook or ignore significant historical processes (Cresswell, 2004; Glassie, 1992).

Like other historical geographies of the present, the assessment of historical significance in CRM must be "calibrated by the shifting sands of interpretation," a fact explicitly accounted for in both the preservation regulations and guidance (Nash and Graham, 2000: 8). After all, like historical geography, CRM is ultimately concerned with the material shaping of places, spaces, landscapes and lives, working at the interplay of the past with the present (Johnson, 2000; Nash, 2000a). Accepting that all interpretation is context bound and power-laden, CRM has embraced the value of linking local knowledge to scientific knowledge in an effort to negotiate the historical geographies of "modernity beyond the metropolis" (Nash and Graham, 2000: 2).

Views of the Public

Traditionally, highway planners and engineers dismissed local geographical knowledge as politically interested and poorly informed, and deployed the weight of their position to overcome local opposition to a project. However, due to long delays associated with organized opposition, some agency officials, particularly those working in rural areas, have embraced the environmental processes and sought to engage the public and plan to incorporate local knowledge into the project development process, believing such engagement furthers—rather than hampers— the development of their undertaking. Indeed, some agency officials have embraced "place-based decision-making', relying on geographical knowledge shaped at the local or regional scale to inform environmental decision-making (CEQ, 1997). By engaging place through consultation with local interests, and discovering the ways in which place-bound genealogies and geographical imaginations (genealogical-geographical knowledge) function to structure local knowledge, unquanitified environmental amenities and values may be given appropriate consideration in the development of specific projects.

The rules and regulations that govern the protection of historic properties place a tremendous emphasis on consultation between agency officials, contractors, Indian tribes, representatives of local governments and the public, including those with a demonstrated interest in or concern with potential effects on historic properties. The National Historic Preservation Act holds that "the views of the public are essential to informed Federal decision-making" and requires agency officials to actively seek and consider the views of the public (36 CFR 800.2(a)4). In fact, the National Historic Preservation Act and the National Environmental Policy Act demand that agency officials regard the views of the public in considering alternatives. Often the public involvement elements of these complimentary laws are blended together during the project development process. Several times throughout the process, agency officials meet with consulting parties to discuss findings and enlist the public in an effort to identify historic properties.

Agency officials share the purpose, goals and timing of the effort to identify historic properties with the public at the outset of the process and solicit information through advertisements, public meetings and meetings with the consulting parties. In an effort to identify potential project constraints, agency officials ask the public to identify significant places and historic sites on constraint mapping, and then commission professional historians to examine the claims in the field and through research. Most of these claims are rooted in the geographical knowledge chronicled, created and reinforced through the work of local historical and genealogical societies.

Some critics have suggested that the incorporation of local knowledge into the environmental process is often biased, exclusive, and captive, used to advance projects at the expense of the public and the cultural environment. Wainwright and Robertson (2003: 201) considered CRM the "work that science performs for the state in its efforts to achieve territoriality," giving a particular interpretation of a place the weight of authority. Claiming that state funded cultural resource assessments are a privileged text put to use as truth and space constituting text "designed to codify and limit dissent," Wainwright and Robertson begin to attack the practices as a colonial project of territoriality, questioning whether it is even possible to write a just cultural resource assessment.

Yet, by situating local histories and genealogies within broader processes operating at wider spatial scales, CRM offers an avenue for practitioners to recover hidden histories, of women, of marginal communities, and of the working class, to recover the genealogies and histories pruned from family trees (Nash and Graham, 2000: 1). The forgotten histories and genealogies excavated in the course of the CRM work, by their recovery and return, are reintroduced to a public that was often unaware of their existence. CRM can bridge the break between how history and genealogy are imagined and how they are actually constituted through material social practices (Harvey, 1993: 16). The local genealogical-geographical knowledge, often viewed as selective, uncritical and authentic, is granted as much weight in the decision-making process as the objective, archival and scientific data (Johnson, 2004: 317-318).

Genealogical-Geographical Imaginations

CRM practitioners have found a need to look beyond the material features of landscapes to examine and account for the personal and subjective meanings of space created through the work of genealogists and propagation of genealogical representations (Lowenthal, 1961; Wright, 1947). The role of the genealogist in the creation of place is linked closely to the perception and imagination of those who share an attachment to a particular geography (Tuan, 1974). The recognition that the genealogical-geographical imagination is compounded by personal experience, learning, imagination, and memory illustrates the care that must be given to the assessment of significance and integrity in the practice of CRM (Lowenthal, 1961: 260).

Further, local history and genealogy are closely tied to this re-creation of power relations, particularly in the non-urban scene. The ability to select, chronicle and perpetuate historical memory has long been a lever at all scales. Even those using local knowledge to oppose a Federal project "must acknowledge the interested and contextual character of local knowledge," but argue that the local perspective, grounded in place, counterbalances the "structured biases built into official knowledge systems which are used by experts to secure employment, control resources, and justify extraction and enclosure" (Robbins, 2004: 120). Although, "for all its promise of certainty, offers no guaranteed solutions to puzzles of belonging and identity," even incomplete genealogies, when tied to place, offer a powerful narrative to support a claim to significance (Nash, 2000b: 1). The linking of genealogical narratives to specific places through the production and dissemination of local histories is an important constituent of a local sense of place.

The uneven geography of genealogical-geographical knowledge is apart when working to identify historic properties in different settings. For example, genealogical information plays a much smaller role in the identification of historic properties during the development of a highway alignment through the dense industrial districts of the Cambria City, Morrellville and Oakhurst sections of Johnstown. This was in part due to the ravages of urban renewal, the extent of previous research on the area's industrial heritage, and immigrant character of the single industry town and the use of history/heritage as a form of social control (Mitchell, 1992). Yet, geographer Sue Friedman discovered a rich collection of genealogical data build up around the Coopersdale section of the city, a spatially discrete section of town that long had an identity as a place set apart from the rest of the industrial ensemble (Heberling et al., 2004). Nonetheless, in urban industrial areas with large ethnic populations, agency officials have found that identity and memory are grounded less in genealogy than in the "old neighborhood", an idealized memory of community (Linkon and Russo, 2002).

Often working in areas hit hard by deindustrialization and resource extraction, CRM can reveal how much history is hidden, how much shadowed ground is absent from both landscape and from historical memory. Revenants, traces of the past kept alive through folklore or local history, are challenging to recognize, let alone to interpret and address within the environmental framework (Mosher, 2005). Yet, even revenants are important factors in the calculus of environmental decision-making as the public brought informal vernacular history and genealogical-geographical knowledge to bear on the formal, supposedly objective, and rational practice of CRM.

Genealogy in Context

Genealogy influences the efforts to identify historic properties both directly and indirectly. Often working within the constraints of a tight budget, occasionally working outside a particular area of expertise, CRM practitioners often rely heavily on the most accessible of materials such as local histories and genealogies found at a local

A Genealogy of Environmental Impact Assessment 69

library or, increasingly, on the Internet. CRM practitioners build on the foundation of the existing information to get to sense of the multiple layers of history that confront them, synthesizing the work of historians, preservationists, and genealogists. Widely disseminated local histories and genealogies inform the development of the historic context, as well as shaping local knowledge and imagination.

The history of Somerset County is well documented in local and county histories and biographies (Blackburn, 1906; Doncaster, 1999; Historical and Genealogical Society of Somerset County, 1980; Walkinshaw, 1939; Waterman, Watkins & Co., 1884). Though popular histories, particularly subscriber-financed histories, are often subjective interpretations of the past, the sum of published histories inform the context and interpretation of specific properties. Local and regional planning documents, architectural histories and historical accounts were shaped to various degrees by the work of genealogists and local historians.

There are several sources of genealogical information on properties within the U.S. 219 study area, most of which are also linked to the study of material culture. The Historical & Genealogical Society of Somerset County's Somerset Historical Center, a state-affiliated museum and 1,000-volume genealogical library, is a repository for an enormous amount of genealogical data. The Center also maintains the Somerset County Historic Sites Survey and Rural Vernacular Architecture Survey—well-funded studies of the regional agriculture and material culture that draw heavily from the genealogical collection (American History Partnership, 1994; David et al., 1993; Somerset County Planning Commission, 1984, 1985a, 1985b; Somerset Vernacular Architecture Survey, 1994). Other materials in the collection, notably a series of oral histories and several historical surveys of coal towns and railroads, explore the history and places of people less likely to have been chronicled in the voluminous subscriber histories and genealogies of the region's "first" families and successful farmer-financiers (Beik, 1996; Mason, 1996; Mulrooney, 1991; Summers et al., 1994).

A second source, and the site of much of the actual production of the genealogical information, is the loose affiliations of historians, genealogists and antiquarians known as the local historical societies. Although smaller and relying on private funding, often working in the shadow of the heavily funded state repository, organizations such as the Berlin Area Historical Society produce, reproduce and maintain an extensive genealogical library, linking their genealogical research to the built environment through programs and popular publications. The Berlin organization has gone so far as to construct, entirely from private funding, a modern genealogical building and library.

Much of the genealogical work generated in the county and local historical societies is captured on the many genealogical websites dealing with the region, including local sites of national web networks (e.g. www.rootsweb.com), state scale sites (e.g. www.pa-roots.org), as well as local sites maintained by families, individuals, churches and the local historical societies themselves. These sites are linked together into a *Somerset County PAGenWeb Project* site, which is a powerful clearinghouse for genealogical information (www.rootsweb.com/~pasomers). Like the local publications, many of the sites link genealogies to specific places or properties.

Genealogical publications, including circulars such as *The Casselman Chronicle* (1961-1981), *Glades Star*, and the *Laurel Messenger* often built genealogies around particular farms or home places. These place-bound genealogies later reproduced in publications such as *'mongst the Hills of Somerset* (Historical and Genealogical Society of Somerset County, 1980), *Exploring Bruedersthal* (Glessner, 2000) and *Legends of the Frosty Sons of Thunder* (Doncaster, 1999), forming a deep local understanding of history and place. Indeed, the local historical societies' work provides both background material and a source of many of the views of the public, whose place perception was shaped by generations of heart-felt and rigorous, if selective, genealogical research.

By assembling a regional context, grounded in a careful examination of the human transformation of the physical environment, the CRM practitioner would ideally develop a closer relationship with the physical and cultural qualities of the locality. This prospect "brings the local landmarks into visibility, giving the creations of a community's people—the artifacts, in which their past is entombed, the texts in their past lives—complete presence" (Glassie, 1982: 621). This presence persists within a constantly changing social, economic and material landscape through the practice of linking genealogy to place.

Research into the history of a property in a region such as Somerset County's Brothers Valley plunges the researcher into the genealogies that swirl around a place. By tracing a chain of title to a farm property, the genealogical relationships within the region become apparent, as the search leads from deeds and mortgages, through probate, wills, orphans' records and tax assessments. When conducted on a large scale in a discrete area, the work reveals the interrelationships between the many historic actors and families as they mediated their relationship to an economy that not only valued the products of their farms, but also the coal below them would draw a transient immigrant population into a small circuit of genealogical relationships. The presence of the immigrant laborers led established families to reinforce their identity, in the landscape through the construction of the massive star barns funded by the sale of mineral rights, as well as through the production of history and place-bound genealogy (McMurry, 2001).

With such a wealth of information, CRM practitioners working in areas such as the Brothers Valley face a very different challenge than those working in areas with a less robust genealogical tradition: that of discerning the validity of strongly held public claims to space that could reflect an incomplete, interested or farcical product of genealogical-geographical knowledge. Given great weight by Federal environmental law and practice, the views of the public and interested parties are important factors in the location and design of a multi-million dollar facility.

'mongst the Hills of Somerset

For the U.S. 219 Improvements Project, the efforts to identify aboveground historic properties are documented in the multivolume Historic Structures Inventory/ Determination of Eligibility and chronicled in an Environmental Impact Statement Report (Hunter and Heberling, 2002; Greenehorne & O'Mara, Inc., 2002).

Following detailed background work, an analysis of census data, comprehensive fieldwork, extensive public involvement and some layering of historic maps, agency officials found that residents have a very strong historical sense of place built on the work of genealogists and distributed through the publication of local historical and genealogical societies. The place-bound genealogical knowledge combined with the empirical results of the field survey link historical memory to the material landscape.

Figure 4.1   The Brothers Valley

In the extractive landscape of Somerset County, Pennsylvania, "capstone landmarks", created of durable materials and well maintained, serve to structure the local geography, and thus reconstitute the sense of place. In Brothers Valley, the scale of this is scarcely believable, whole communities and landscapes physically erased by creative destruction and resource extraction (Figure 4.1). Over generations, the less durable vernacular environment most often associated with immigrant labor is no longer visible, and a landscape is created and then reinforced that is divorced from the complete history of the place. Over time, place-bound genealogies assign tremendous significance to the capstone features in this residual landscape (Figure 4.2).

In spite of the landscape's duplicity and uneven genealogical-geographical terrain, the *Bruedersthal* is a place deeply affected by its history. Many of the first post-pioneer settlers, the initial occupants that exert a cumulative influence over subsequent cultural landscapes, were from Germany, eastern Pennsylvania and western Maryland (Kniffen, 1965). Some were Scotch-Irish immigrants from Ulster,

Figure 4.2   Genealogy in place: The Walker Farmstead

or migrants of Scots-Irish descent, arriving from established hearths along the seaboard. However, migrants of German heritage were dominant at the time of the first effective settlement, an influence that resonates in the cumulative landscape of the Brothers Valley region (Berg, 1999; Kniffen, 1965).

The *Bruedersthal* developed as a center of dairying and maple sugar production, though farming was far from specialized. Local farmers practiced a form of mixed agriculture classified as general farming: growing grains and raising livestock for the market in addition to growing fruit, grapes, and vegetables as supplemental crops. As the Brothers Valley community entered the last quarter of the nineteenth century, prosperous farmers bought, bred, and sold cows for beef and dairying; produced large quantities of butter; raised sheep that grazed the upland portions of the farm properties and yielded wool to supply the regionally strong textile industry. Nearly all farmers raised small stock such as pigs and fowl to provide eggs.

Sometimes, the second and third generation would inherit a substantial farm of considerable value, often with a modest if sturdy open-forebay standard Pennsylvania bank barn, the first house built on the property, and a washhouse. In contrast to the primogeniture frequently practiced by Scots-Irish Upland Southerners, the farmers of the Brothers Valley distributed land among their family, subdividing large tracts and acquiring adjacent farms, resulting in clusters of actually and thematically related farms. Once land was subdivided, the sale of coal rights and timber would often finance the reconstruction or construction of a new residence, frequently involving the incorporation of the traditional plan of the parent farm or reuse of original materials.

The agricultural landscapes of the project area reflect a traditional practice of market agriculture, particularly the production of maple syrup and dairy products. This agrarian landscape is overlain by the industrial era landscapes of coal, limestone, and timber extraction. Capital from the sale of mineral rights enabled local farmers to construct modern agricultural facilities and maintain modest farms that now structure the local sense of place and identity, the sugar camps and star barns bound to place by complex and interwoven genealogical narratives. The residual landscapes associated with specific types of agriculture, mining, manufacturing and transportation combine to form a layered landscape, in which the boundaries between farms and industrial landscapes are continually defined and redefined (McMurry, 2001: 135). In the face of large-scale environmental transformation, regional identity is located in the place-bound genealogies built around the capstone features in the landscape.

Rural genealogical projects involve a claim to space that has been challenged through economic and social restructuring. Landmark houses and massive "star" barns loom above the coal patch towns once densely populated with immigrant laborers; the genealogical project to claim the farms buttress this material imprint of hierarchy and hide the life worlds of the very people that produced the wealth materialized in these barns so large they could never be filled. Contrast the place-bound protective genealogies of central Pennsylvania, grounded in places that contain little or no resonance of the ethnic workers who generated the wealth materialized in the landscape, with the quest of the worker's descendents to find a placeless home in an ethnic affinity. One genealogical narrative built around movement, the other rooted in place.

In rural areas such as Somerset County, the role of place-bound genealogy has to be considered in light of the changing social and economic fortunes of the long time residents. Deindustrialization, changes in the nature of agriculture and the ravages of mineral extraction have led to extensive out migration from the region, signaling a loss of local control over economic fortunes. Once mediating their position within the world economy on strong terms, the descendents of once powerful farmer-financiers have left the region or have entrenched themselves in the ruins of the regional economy through the construction and reinforcement of very selective place-bound genealogies.

Genealogy in Place

Although the majority of the people who lived and worked in the Brothers Valley through history were not farmers or even landowners, farm families and their descendents are most active in creating and disseminating place-bound genealogies. Indeed, folklorist Cynthia Mason, exploring the intersection of the agricultural and industrial communities, concludes that there is not a single unified story to the region's history, particularly in light of the brutal suppression of organized labor in the Brothers Valley (Mason, 1996). Yet, traditionally agricultural families that were able to maintain their economic position came to have a particularly sustained agency over the production of the landscape. The dense and well-documented genealogies link these families to specific places, highlight the significance of an assemblage of properties, and create a strong current of continuity that persists in landscape and comes to dominate place perception (Hunter and Heberling, 2002; McMurry, 1994).

For example, the Walker family occupies a central role in the history and development in this portion of Somerset County (Hunter and Heberling, 2002). From settlement to the 1970s, the Walker family developed a mosaic of agricultural and commercial enterprises in the region surrounding this family seat. Additionally, the work of genealogists ensures that the family patriarch, Jacob Walker, enjoys near folk hero status in the Brothers Valley area. Although by some accounts, Jacob Walker was a native of Germany who had emigrated prior to the Revolution, other sources suggest he was born in Fredrick County, Maryland in 1745, the son of Francis Walker, an immigrant from England by way of Holland. By 1774, Walker relocated north into the Brothers Valley, establishing a farm, though retaining his property in Maryland. Walker served as an officer in the militia during the Revolution, but did not live to see the articles of peace ratified; he perished following a riding accident, an incident that is central to the historical narrative of the Brothers Valley.

As the local genealogies have it, in the autumn (some say the spring) of 1783, the settlers in the lower Brothers Valley were intent on establishing a town of consequence on the ridge dividing what is now Swamp Creek and Blue Lick Creek. The legend has it that a horse race was run up the Swamp Creek Valley, the winner selecting the first lot of the new town near what is now Pine Hill. Jacob Walker is said to have nearly achieved victory when his horse lurched at a fork in the trail and he was thrown against a tree and critically injured. Walker is buried in a small plot on the adjacent farm.

Figure 4.3 **The Swamp Creek Valley facing south**

Both the nearby gravesite and what is said to be the tree, located in the upper portion of the Swamp Creek drainage, are significant local landmarks that serve as bookends to the lower Brothers Valley area (Figure 4.3). Many local residents are emphatic that it is in fact where Walker met his end, and assigned tremendous significance to the site, in spite of the apparent youth of the actual tree. Walker's will states that the land was to be distributed into three parcels for his three sons, the lower portion given to Philip, the middle to Jacob, and the upper to Peter. The Walker family expanded its holdings in the region throughout the following decades, owning no less than ten properties in northern Summit and southern Brothers Valley townships.

The genealogies revealed that the third generation, farmers such as Hiram Peter Walker, inherited their properties on the cusp of the dramatic economic transformation of the Brothers Valley area. Known for his business acumen, Hiram P. Walker thrived in a wholly new economic landscape than that of his father. By 1876, regional coal reserves had began to be exploited in a systematic manner with the development of the railroad. The Hiram P. Walker property was located in between towns and mines, at the center of a constellation of Walker family holdings.

Hiram and Elizabeth Walker heralded the property into the twentieth century, selling the 363-acre farm to their fifth son, Frank E. Walker in January 1909 (Somerset County Deed Book, 161: 215). The relationship between local farmers and coal mining is made explicit in the 1909 deed transferring the property to Frank E. Walker. The deed explains that the grantors, Hiram and Elizabeth Walker, reserve the conditions for the exploitation that the coal seam known as the "L" prime vein, and all coal beneath the vein. Exempt from the transfer is a tract of land at the western boundary of the property, reserved for the purpose of "erecting the necessary mine buildings, mine shafts, or slopes and other mine operations necessary to mine and remove coal". The deed explains the requirements for a fenced road, airshaft, underground drainage pipes, and the protection of the farm buildings, and that the reservation is to revert to the owner of the land "when all coal from the farm is removed" (Somerset County Deeds 38, 411-412; 161, 215).

The next decades marked perhaps the most dramatic reordering of the local circulation network since the coming of the railroad. Technological advances allowed for the development of the fluid milk industry in the region, reflected in the reorientation of barns, fields and support structures. Perhaps due to the public works developed during the Depression, or as part of the movement for good roads born out of the developing dairy industry, the road network was dramatically reordered. The main east-west route shifted away from the buildings to the current configuration, and the old roads faded into field lines and traces.

In the subsequent years, the owners and administrators of the farm continued to invest in new buildings and equipment. Yet, the required reinvestment made competition increasingly difficult, the sale of surface mining rights and timber was not enough to compete in an ever-tightening market, and the family slowly backed away from commercial agriculture. The mechanization of agriculture, economies of scale, the lack of off-farm economic opportunity and the lack of access led to a dramatic out-migration and an atrophying of the rural landscape. On 12 October 1970, executors passed the remainder of the farm out of the Walker name for the first time since settlement (Somerset County Deed Book 691, 292, 429).

Genealogical Production of Landscape

The results of the cultural resource survey and public involvement were clear: a cluster of related properties within a small watershed was an intact remnant of the once expansive Brothers Valley region, the *Bruedersthal*, associated with a range of significant events, persons, and movements in architecture. Swamp Creek drainage, once known as the Pine Run, was one of the few portions of Brothers Valley to have escaped extensive surface mining at the time of the survey, a remnant of what once was a much larger landscape. The well-defined area consists of 15 farms (four individually eligible for the National Register) and two cemeteries, many developed by the Fritz and Walker families (Figure 4.4).

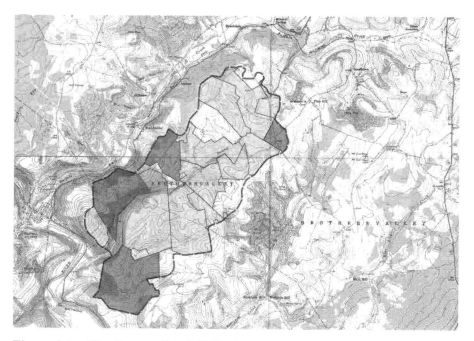

Figure 4.4 The Swamp Creek Valley historic district

According to the National Register guidelines, a historic district is a geographically definable area possessing a significant concentration, linkage, or continuity of sites, buildings, structures, and/or objects united by past events or aesthetically by plan or physical development (National Park Service, 1998). Districts are historic environments that convey a sense of time or place through the survival of interrelated resources that represent important themes of the area's historic development. A district may represent an important period or feature in the history of an area, or it may illustrate change over time. Eleven characteristics are common to rural historic landscapes: evidence of land uses and activities; patterns of spatial organization; response to the natural environment; cultural traditions; circulation networks; boundary demarcations; vegetation related to land use; buildings, structures and objects; clusters of features; archaeological sites; and small scale elements.

The sum of these resources, in addition to other material elements, appears to constitute "a significant concentration ... [and] continuity of areas of land use, vegetation, buildings and structures, roads and waterways, and natural features," the characteristics of a rural historic landscape. Further, the landscape clearly reveals "a rural community whose organization, buildings and structures, or patterns of land use reflect cultural traditions valued by its long-term residents," the definition of a traditional cultural property (King, 2000: 176). The Pennsylvania State Historic Preservation Office formally determined the Swamp Creek Valley Historic District eligible for the National Register of Historic Places on 26 September 2002. The designation of the Swamp Creek Valley as a historic district insures the long-term active management of the landscape during all federally funded or permitted undertakings within its boundaries.

Several properties retain a high degree of integrity, and are individually eligible for the National Register of Historic Places. Old road alignments, traces, and building orientation map the changes in circulation and production. The fence lines and field patterns appear to reflect early property boundaries and important subdivisions. Intrusions such as the surface coal mines, quarries, and modern homes are enveloped by the sum of the significant physical and cultural features, and are effectively lost in the landscape. Owing to its declining fortunes, the district has enough important material features to create an unusually picturesque tableau: the un-stripped fertile valley framed by high ridges.

From the beginning of the project, agency officials were aware that, due to the history and character of the region, historic properties would be affected by the undertaking, requiring, under Section 106, some form of mitigation. The size and extent of the Swamp Creek Valley Historic District ensured that all build alternatives would use land from and adversely affect the district, requiring ongoing consultation and development of mitigation strategies. As a result, agency officials decided at an early stage to focus on minimization of the project effect and mitigation through consultation with the public. Given the size of the district and likelihood of adverse effects, the place-based mitigation measures were chronicled in a programmatic agreement between the Federal Highway Administration and the Pennsylvania State Historic Preservation Officer, with the assent of the Pennsylvania Department of Transportation and the Advisory Council on Historic Preservation.

During the final design process, Federal Highway Administration committed to ensure that the section of new highway located within the boundaries of the Swamp Creek Valley Historic District incorporates design features that will minimize physical and visual effects on contributing properties in the historic district while meeting current Federal and state highway design standards. These design features include the use of form liners and stained concrete to simulate natural stone on the abutments, wing walls, parapets and piers of the proposed bridges over Swamp Creek and Pine Hill Road; the preservation of a riparian corridor along the streams spanned by the new highway; use of the steepest cut and fill slopes permissible under Federal and state design standards to minimize right-of-way acquisition; and reduction of median and shoulder widths to the minimum required by current design standards. The Federal Highway Administration also committed to ensure that the edges of the highway right-of-way are landscaped to provide a visual and audible buffer between the highway and district in accordance with a landscaping plan developed during ongoing consultation.

In testament to the importance of place-bound genealogies to the development of the project, the Federal Highway Administration committed to ensure a Public Mitigation Advisory Committee to assist in the development of public education materials appropriate to the resource during the final design process, and to consult on future undertakings within the district. The Public Mitigation Advisory Committee consists of designated representatives of the local historical societies, as well as agency officials. The Public Mitigation Advisory Committee will continue to inform the development of any future federally funded or permitted undertaking within the district, highlighting the dialectical relationship between abstract genealogical-geographical knowledge and the material landscape.

Conclusions

Following thirty years of environmental impact assessment, environmental professionals are finally embracing place-centered strategies in the practice of cultural resource management. CRM has the potential to be an essentially moral project: a participatory, rigorous and non-coercive practice that allows genealogical-geographical knowledge to fill a gap in environmental impact assessment (King, 1998; Sack, 1997). Through the cautious incorporation of, interrogation of and engagement with the genealogical-geographical imagination, CRM can not only protect the significant artifacts of the past, but produce a future landscape rich in meaning, symbolism, and local identity.

References

American History Partnership (1994), *Somerset Historical Center Interpretive Manual*, Youngstown: American History Partnership.
Anderson, B. (1998), "The Importance of Cultural Meaning in Defining and Preserving a Sense of Place", in Tomlan, M.A. (ed.), *The Preservation of What, for Whom? A Critical Look at Historical Significance*, Ithaca, NY: National Council for Preservation Education, pp. 127-135.
Beik, M. (1996), *The Miners of Windber: Struggles of the New Immigrants for Unionization, 1890s-1930*, University Park: The Pennsylvania State University Press.
Berg, L. (1999), *Somerset Historical Center Exhibit Research: Context Summaries*, Somerset, PA: Interpretive Solutions-Somerset Historical Center.Blackburn, E. (1906), *History of Bedford and Somerset Counties, Pennsylvania*, Chicago: The Lewis Publishing Company.

Bodnar, J. (ed.) (1992), *Remaking America: Public Memory, Commemoration, and Patriotism in the Twentieth Century*, Princeton: Princeton University Press.
Casselman Chronicle (1961-1981), *The Casselman Chronicle Volume 1 to Volume 21*, Springs: Springs Historical Society of Casselman Valley.
Council on Environmental Quality (CEQ) (1997), *The National Environmental Policy Act: A Study of Its Effectiveness after Twenty-five Years*, Washington: Executive Office of the President.
Cresswell, T. (2004), *Place: A Short Introduction*, Malden, MA: Blackwell.
David, L., Buddle, B. and Mason, C. (1993), *Rural Vernacular Architecture Survey Phase I: Brothers Valley and Lower Turkeyfoot Townships*, Somerset: Somerset County Historical Center.
Doncaster, W. (1999), *Legends from the Frosty Sons of Thunder*, White Stone: Bradylane Publishers.
Duncan J., Johnson, N. and Schein, R. (eds) (2004), *Companion to Cultural Geography*, Malden, MA: Blackwell.
Glassie, H. (1982), *Passing the Time in Ballymenone*, Bloomington: Indiana University Press.
Glassie, H. (1992), "Artifact and Culture, Architecture and Society", in Bronner, S. (ed.), *American Material Culture and Folklife: A Prologue and Dialogue*, Logan, UT: Utah State University Press. pp. 47-62.
Glessner, C. (2000), *Exploring Bruedersthal: A Traveler's Guide to the Historical Berlin Area*, Berlin: Glessner & Shumaker.
Graham B.J. and Nash, C. (eds.) (2000), *Modern Historical Geographies*, Harlow: Pearson.
Greenhorne & O'Mara, Inc. (2002), *U.S. 219 Improvements/S.R. 6219, Section 020: Environmental Impact Statement and Preliminary Engineering, Somerset County*. Hollidaysburg, PA: Pennsylvania Department of Transportation District 9-0.
Groth, P. (1997), *Vision, Culture, and Landscape*, Berkeley: University of California, Berkeley, Department of Landscape Architecture.
Harvey, D. (1979), "Monument and Myth", *Annals of the Association of American Geographers*, 69: 362-381.
Harvey, D. (1993), "From Space to Place and Back Again: Reflections on the Condition of Postmodernity", in Bird, J., Curtis, B., Putnam, T., Robertson, G. and Tickner, L. (eds), *Mapping the Futures: Local Cultures, Global Change*, London: Routledge, pp. 4-29.
Heberling, S.D., Hunter, W.M. and Friedman, S.W. (2004), *West End Transportation Improvements Project, S.R. 0056, Section 023, Cambria County, Pennsylvania Volumes I, II and* III, Alexandria: Heberling Associates, Inc.
Historical and Genealogical Society of Somerset County (1980), *'mongst the Hills of Somerset': A Collection of Historical Sketches and Family Histories,* Somerset, PA: Historical and Genealogical Society of Somerset County.
Hobsbawm, E. (1983), "Mass Producing Traditions: Europe, 1870-1914", in Hobsbawm, E. and Ranger, T. (eds), *The Invention of Tradition*, Cambridge: Cambridge University Press, pp. 263-307.
Hobsbawm, E. and Ranger, T. (1983), *The Invention of Tradition,* Cambridge: Cambridge University Press.

Holdsworth, D. (1997), "Landscape and Archives as Texts", in P. Groth and T.W. Bressi, (eds), *Understanding Ordinary Landscapes*, New Haven: Yale University Press, pp. 44-55.

Hunter, W.M. and Heberling, S.D. (2002), *U.S. 219 Improvements Project, S.R. 6219, Section 020. Historic Structures Inventory/Determination of Eligibility Report Volume I, II and* III, Alexandria: Heberling Associates, Inc.

Johnson, N.C. (2000), "Historical Geographies of the Present", in Graham, B.J.and Nash, C. (eds), *Modern Historical Geographies*, Harlow: Pearson, pp. 251-272.

Johnson, N.C. (2004), "Public Memory", in Duncan, J.S., Johnson, N.C. and Schein, R.H. (eds), *A Companion to Cultural Geography*, Malden, MA: Blackwell, pp. 316-328.

King, T. (1998), "How the Archeologists Stole Culture: A Gap in American Environmental Impact Assessment Practice and How to Fill It", *Environmental Impact Assessment Review*, 18: 117-133.

King, T. (2000), *Federal Planning and Historic Places: the Section 106 Process*, New York: Alta Mira Press.

Kniffen, F. (1965), "Folk Housing: Key to Diffusion", in Upton, D. and Vlach, J.M. (eds), *Common Places: Readings in American Vernacular Architecture*, Athens, GA: University of Georgia Press, pp. 3-26.

Linkon, S.L. and Russo, J. (2002), *Steeltown U.S.A.: Work and Memory in Youngstown*, Lawrence: University of Kansas Press.

Lowenthal, D. (1961), "Geography, Experience, and Imagination: Towards a Geographical Epistemology", *Annals of the Association of American Geographers*, 51: 241-260.

Lowenthal, D. (1977), "Past Time, Present Time: Landscape and Memory", *Geographical Review*, 65: 1-36.

Mason, C. (1996), *Farming and Mining Communities: A Collective Memoir*, Somerset, PA: Folklife Documentation Center, Somerset Historical Center.

McMurry, S. (1994), *Agriculture and Vernacular Building in Somerset County: A Story of Tradition and Change*, Somerset, PA: Somerset Historical Center.

McMurry, S. (2001), *From Sugar Camp to Star Barn: Rural Life in a Western Pennsylvania Community, 1780-1940*, University Park: Pennsylvania State University Press.

Merrifield, A. and Swyngedouw, E. (1997), *The Urbanization of Injustice*, New York: New York University Press.

Meethan, K. (2004), "'To Stand in the Shoes of my Ancestors': Tourism and Genealogy", in Coles, T. and Timothy, D.J. (eds), *Tourism, Diasporas and Space*, London: Routledge, pp. 139-150.

Mitchell, D. (1992), "Heritage, Landscape, and the Production of Community: Consensus History and its Alternatives in Johnstown, Pennsylvania", *Pennsylvania History*, 59: 198-226.

Mosher, A. (2006), *Low Bridge, No Bridge: Public Memory and Creative Destruction Along the Erie Canal*, Paper presented at the Annual Meeting of the Association of American Geographers, Chicago, 11 March.

Mulrooney, M. (1991), *A Legacy of Coal: The Coal Company Towns of Southwestern Pennsylvania,* Washington, DC: Historic American Buildings Survey/Historic American Engineering Record.
Nash, C. (2000a), "The Historical Geographies of Modernity", in Graham, B.J. and. Nash, C. (eds), *Modern Historical Geographies*, Harlow: Pearson, pp. 10-40.
Nash, C. (2000b), "*Genealogical Identities*", Paper presented at the Institute of Advanced Studies, The University of Western Australia, July 2000.
Nash, C and B. Graham. (2000), "The Making of Modern Historical Geographies", in Graham, B.J. and. Nash, C. (eds), *Modern Historical Geographies*, Harlow: Pearson, pp.1-9.
National Park Service. (1996), *How to Apply the National Register Criteria for Evaluation: National Register Bulletin 15,* Washington, DC: National Park Service, Interagency Resources Division.
National Park Service. (1998), *National Park Service Guidelines for Evaluating and Documenting Rural Historic Landscapes: National Register Bulletin 30*, Washington, DC: National Park Service, Interagency Resources Division.
Osborne, B. (1998), "Constructing Landscapes of Power; the George Etienne Cartier Monument, Montreal", *Journal of Historical Geography*, 24(4): 431-458.
Peet, R. (1996), "A Sign Taken for History: Daniel Shay's Memorial in Petersham, Massachusetts", *Annals of the Association of American Geographers*, 86(1): 21-43.
Robbins, P. (2004), *Political Ecology: A Critical Introduction*, Malden, MA: Blackwell.
Sack, R. (1997), *Homo Geographicus: A Framework for Action, Awareness, and Moral Concern*, Baltimore: The Johns Hopkins University Press.
Secretary of the Interior (1983), "The Secretary of the Interior's Standard and Guidelines for Archaeology and Historic Preservation", *Federal Register*, 48: 190.
Soja, E. (1997), "Planning in/for Postmodernity", in Benko G. and Strohmayer, U. (eds), *Space and Social Theory: Interpreting Modernity and Post Modernity*, Oxford: Blackwell, pp. 236-249.
Somerset County Planning Commission (1984), *Preliminary Report: Somerset County Historic Resource Survey*, Harrisburg: Somerset County Planning Commission.
Somerset County Planning Commission (1985a), *Preliminary Report: Somerset County Historic Resource Survey*, Harrisburg: Somerset County Planning Commission.
Somerset County Planning Commission (1985b), *Data Analysis: Somerset County Historic Resource Survey*, Harrisburg: Somerset County Planning Commission.
Somerset Vernacular Architecture Survey (1994), *Rural Vernacular Architecture Study*. Appendix I, Somerset: Somerset County Historical Center.
Summers, P. Rose and G. Fitzsimons. (1994), *Somerset County, Pennsylvania: An Inventory of Historic Engineering and Industrial Sites*, Washington, DC: Historic American Buildings Survey/Historic American Engineering Record and America's Industrial Heritage Project, National Park Service.
Tuan, Y. (1974), *Topophilia: A Study of Environmental Perception, Attitudes, and Values*, Englewood Cliffs: Prentice-Hall.

Tuan, Y. (1977), *Space and Place: The Perspective of Experience*, Minneapolis: University of Minnesota Press.

Walkinshaw, L. (1939), *Annals of Southwestern Pennsylvania Volume IV*, New York: The Lewis Publishing Company.

Wainwright, J. and Robertson, M. (2003), "Territorialization, Science and the Colonial State: the Case of Highway 55 in Minnesota", *Cultural Geographies*, 10: 196-217.

Waterman, Watkins and Company (1884), *History of Bedford, Somerset, and Fulton Counties*, Pennsylvania, Chicago: Waterman, Watkins, and Company.

Wright, J. (1947) "Terre Incognitae: The Place of Imagination in Geography", *Annals of the Association of American Geographers*, 37: 1-15.

Zukin, S. (1997), "Cultural Strategies of Economic Development and the Hegemony of Vision", in Merrifield, A. and Swyngedouw, E. (eds), *The Urbanization of Injustice*, New York: New York University Press, pp. 223-243.

Chapter 5

Knitting the Transatlantic Bond: One Woman's Letters to America, 1860-1910

Penny L. Richards

Introduction

In 1852, John Glencross (1821-1894) and his young wife, Helen Brown (1830-1855), left the Bogg, a farm near Sanquhar, Dumfriesshire, Scotland, for America. The couple settled in Dunmore, Pennsylvania, where their daughter Marion Glencross (1852-1919) was soon born. By 1855, Helen was dead, and John was raising Marion on his own, running a small farm and working in the local coal mines. John Glencross never returned to Scotland, and his daughter never visited her parents' home country, though she would marry James Bryden (1845-1895), a later Scottish immigrant who also worked at the mines. Eventually, John's brother Joseph (1819-1898) would also come from Dumfriesshire, to live nearby, in the anthracite region of Pennsylvania. From these bare facts, it might appear as if the Glencrosses made a "clean break" with the home country. But other evidence tells a subtler story of their immigration. That evidence comes from their Dunmore house, but not from their own words. Instead, we learn of their continuing relationships in Scotland from the writings of a kinswoman that Marion Glencross Bryden never met: another Marion, Marion Brown (1843-1915). Marion Brown's letters to John (and later to Marion Glencross and James Bryden) would be passed from granddaughter to granddaughter until a century had passed, and their contents were again read, for evidence of the bonds they reflected, created, and maintained.

Knitting the Bond

> I was begun to knit a pair of black and white stockings to Marion to let her see some of our Sanquhar patterns ...
>
> Marion Brown to John Glencross, 20 September 1869

Marion Brown could knit. In the long periods of her life when she was too ill to stand, or even sit up, she would knit stockings, sweater vests, hats, and other useful items for her family. *Knitting* had a specific meaning in Sanquhar, Dumfriesshire, Scotland, where Marion lived most of her life. Wool production was a strong component of the local economy. Traditional Sanquhar knitting involves an intricate graphic pattern of contrasting squares. Sanquhar-knit gloves are distinctive for the way the fingers and thumb are constructed for an unusually good fit, and have long been considered a symbol of Sanquhar folk culture. At the 1886 International Exhibition of Industry, Science and Art in Edinburgh, Sanquhar gloves were featured as one of the representative traditional crafts in the section labeled "Women's Industries" (Dumfries and Galloway Museums Service, 1998).

Family, too, had a specific meaning for Marion. At its core, Marion's family included John Glencross's sister, Agnes Glencross Scott, the aunt who raised her and cared for her into adulthood; and Thomas ("Tam") Scott, Agnes's son, the cousin who served as male head of their household. But it also included overseas family members, cousins she had never met in Pennsylvania's anthracite region, cousins like Marion Glencross Bryden. In this understanding of family, she was much like other residents of Sanquhar, a town which lost more than a quarter of its residents in the second half of the nineteenth century, mostly to out-migration, either to other locations in Great Britain, or overseas (Bartholemew, 1895; Brown, 1891). Family, like knitting, was another project for Marion, one that also involved the skillful gathering of strands into something complex, something ideally comforting and durable.

Marion Brown knitted the family bond across distance, decades and generations. Literally, she sent knitting to the transatlantic cousins, as the opening quote of this section suggests, but more critically, she maintained a correspondence over at least forty years with them. She needed their sympathy, and as the years went on, she also needed their material support. In return, she tried to present herself as a useful homeland correspondent—a source of family and Sanquhar news, an advisor on feminine culture for motherless Marion Glencross, and a possible emigrant herself someday.

Marion Brown's actual knitting has probably not survived. Wool becomes moth-eaten, blankets grow threadbare, and newer garments replace older ones. Even if a scrap still existed, it would not be labeled or signed as her work. But her letters, about 150 of them, were signed and have survived, remaining at the Pennsylvania address where she sent them, in the hands of her addressees' female descendants. Several of the photographs she sent can also still be found in family albums, among other sepia-toned studio portraits of solemn Scots. (There are three known images of Marion Brown herself.) Her woolen designs perished with the years, but much of her writing remains, to testify to this one woman's decades of work at making and keeping family bonds across time and ocean.

Especially when exchanged among far-flung relations, family letters are an excellent data source for both family historians and academic researchers interested in the history of immigration. The Marion Brown letters can be read in both contexts. This collection points to the role of women as family historians, information recorders and gatekeepers. They also represent a rare genre in several respects: as letters from the family members who did *not* emigrate; as the letters of a rural Scottish woman; and as a firsthand account of living an ordinary life with a disability (Gerber, 1997; Breitenbach et al., 1998; Bredberg, 1999). Finally, their survival in private hands raises questions about the preservation of such resources, and outreach to find more such treasures before they disintegrate.

Reading Correspondence

Letters are seemingly intuitive documents—most adults today have read and written letters, so we feel confident that we understand their conventions, their circumstances. As private communications, they have an unedited, unpolished quality that appeals to the historian's interest in the "real" story, unfiltered through memory or formality. For family historians, letters are also prized because they hold actual handwriting and words directly from ancestors themselves; touching an old letter means touching the same paper and ink that great-grandparent touched, and there is something powerful about that physical connection for many family historians. For researchers interested in histories of a broader sort, correspondence is a key source too. While quantitative history can reveal much about the aggregate experience of the past, the individual experience is still best read in words. The academic historian reads correspondence somewhat more critically, looking at the "epistolary discourse" being employed and asking questions about reliability and representation. Letters are not necessarily as transparent as they first appear.

Letters and Family History

When reading letters written by ancestors, family historians find a "small treasure" of information (Jolly, 2005). First, letters contain details of births, deaths, marriages, and other names, dates, and places that form the skeleton of a family history project. Especially when letters are written between distant correspondents, such news will often form the bulk of the letter, and cover not only immediate family but neighbors, distant relations, and others. This information is in most cases specific and detailed, perfect for a genealogical application. When later memories fail or falter, and when official documents are missing or conflict with family lore, letters written "on the scene" stand as more reliable records of when and where life events took place. Beyond these data, family letters are prized as something unusually "real", "true", and close to the source. "Nothing tells the true reality of war more than the simple writings of the common soldier," explains one genealogical site dedicated to sharing private correspondence among family history researchers (Schulze, 2002: np). These letters provide us with a sense of history, of being there and experiencing life with the people who write. Many family historians confess to the hope of reading a secret, of gaining a poignant or humorous anecdote, in their work with private correspondence. Finally, letters can help identify sitters in unlabeled portraits, by mentioning the photograph being made or narrating its contents.

Marion Brown's letters work on all these levels for the family historian. Though her range was small—she rarely left the family home, because of complicated health problems that kept her often in bed—she was a careful reporter of all the life events in her midst; she notes the birth of every one of Tam Scott's eleven children, along with their names and sometimes an explanation of their names:

> We have got another addition to our family Robina had a son on the 5th of this month I am very glad to tell you they are both doing well. And this baby has to be named for the doctor it is the fashion here when a new doctor comes the first baby he brings home is named for

him so it happened that Robina was the first in Sanquhar and his name is George Sinclair Scott. (Brown to M. Glencross Bryden, 19 December 1888)

Deaths in the household are also described in full detail, on black-edged mourning stationery. Marion Brown's first separate letter to Dunmore was one such sad report:

> I had little thought that the first letter I was to write to you was to inform you of Uncle James's death for although he was not very strong we little expected we was to lose him so soon[;] he was not very strong the whole spring and he took a bad cold about two months since and he was just getting a little better when he took a belling throat and after that he had congestion of the brain which is a very dangerous trouble and the Docters said it proceeded from a sore ear he has had for a long time he was confined to his bed for three weeks and I think the most of his pain was past before he took the bed for he never complained of anything scarcly... (Brown to J. Glencross, 25 July 1866)

A "belling", or bealing, describes swelling, infection, and abscesses—it is an obscure Scots term still sometimes found in parts of eastern Canada, and Appalachia, as well, according to the *Dictionary of American Regional English* (Cassidy 1985). To compound the sadness, the very next dated letter in the collection reports another death in the family. And here is how Marion explained Tam Scott's change in marital status in 1879:

> Tom is going to be married on the last day of this month if all goes well. The woman his is going to marry belongs to New Cumnock her name is Robina Boyle he is going to have no wedding only a marriage she is coming here to live along with us. (Brown to J. Glencross, 23 January 1879)

We get all the basic announcement information here: bride's name, bride's hometown, date of the union, and intended living arrangements. The comment "he is going to have no wedding only a marriage" points to Scott and Boyle contracting an "irregular marriage", without any church or civil ceremony—nothing so quaint as a handfasting or other informal rituals, an irregular marriage generally meant no ceremony or publication at all—so only a private document such as this letter would ever record its date. Such irregular marriages were legal in Scotland until 1940 (Leneman, 1999).

Brown's letters help make connections otherwise hard to discern. When John Glencross moved to America, he was married to Helen Brown, and they had their daughter Marion Glencross in America. At a glance, theirs seems like a small but simple nuclear family pattern, but through the letters, we discover that John Glencross had an older son (called John Glencross, born 1844) by a woman named Christian Neilson, and that the boy and his mother were still alive when John left Scotland in 1852. The whereabouts and well-being of the younger John Glencross are frequently included as news in Marion Brown's letters to the American Glencrosses, and at least once she sent along a photograph of the young man ("a strapping good-looking young man", in Marion's judgment, in a letter dated 7 April 1869). Another example can be noted in that when Helen Brown died, she was only 25, and there are no further letters from her relatives. Their story might be absent from the family tales, but Marion Brown made reference to Helen's family, tracking down news for John and Marion Glencross's sakes:

> I have found out Helen's mother at last I got a letter from her the other day and she is well and very comfortable she is housekeeper to a lady and just goes about with her it seems they are not long in one place but now when I have found her out I will can send her the money to the address she has sent me and she would like very well if you would send your likeness and Marion's both on one card and you are to send the card to me as I cannot tell you how long she may be at the same place... (Brown to J. Glencross, 6 April 1870)

Marion Brown thus managed information for a family full of tangled, loose, and cut threads, reattaching granddaughter and grandmother years past when death, immigration, and itinerant work might have severed the connection. She gathered news of an estranged son for his father, and hoped to convey to a traveling housekeeper some images and support from their common relations.

Brown's letters are certainly full of hard truths. Marion Brown was disabled. At various times in her life, her mobility, sight, and speech were impaired for long stretches, and she spent months in bed, unable to speak or even sit up (Hutchison, 2003). We cannot at this distance discover the cause of her disability, nor the name it would be given today; we only know what symptoms and restrictions she recorded, and their combination does not suggest any one clear diagnosis. She endured medical treatments involving "galvanism", head shaving, blisters, leeches, and other measures of questionable purpose. She lived in a small house with a lot of other people (various older uncles early on, then Agnes Scott, Tam Scott, Robina Scott, and eventually their eleven children). Money was scarce, her aunt and cousin were in poor health themselves, and the family moved often with changing circumstances. Tam Scott found himself jailed briefly for assault in 1873, further affecting his health and earnings. A constant thread in the letters is how difficult it was to farm at "the Bogg":

> [T]here is not many diseased potatoes but they are very thin in the ground and for turnips Aunt has none this year they are all rotten with what they call finger & toe I don't know whither the master will allow her anything for the bad crop or not. The cows is all very healthy now but they have never given the same quantity of milk the whole summer as they did before they had the trouble. (Brown to J. Bryden, 10 October 1872)

Agnes Scott was a cheese maker of some local repute, so any dip in milk production would have affected her livelihood. Later, we get a glimpse of what the emotional atmosphere was like for Marion:

> Now James if you was in my place how would you like if any on[e] was to say to you that you ought to be in the poor house one day when Tam's wife said that tie me... neither Tam nor Aunt heard her say it to me nor would I not tell them for it would do nothing but vex them both and Tam would give the last halfpenny he had for either of us and they have both enough to think of without me telling them anything... I don't want you to say a word about what I have told you but my heart was sore when I felt my own helplessness and what a burden some of my friends think I am. (Brown to J. Bryden, 5 September 1881)

In time, despite her significant health issues, Marion Brown left the Scott household, with all its pressures. She wrote from the telephone office in Sanquhar in the late 1890s, saying

> it would have been far better for us both if Aunt would have come to a house by ourselves when I wanted her, but as far as I can see she will have to do where she is now, and if she is taken from the world before me, as we don't know who will go first, I will be left without a home... (Brown to M. Glencross Bryden, 9 January 1896)

It would hardly be possible for a descendant of Brown's correspondents to read these letters and come away with a romanticized picture of the life left behind when John and Helen Glencross left Scotland for America in 1852. Helen's mother was working as a housekeeper with no fixed location, we learn. John's family was crowded into a farmhouse, his male cousin a struggling laborer who worked at a brickworks or on road-building projects when the leadmines were not hiring. Marion's own pain, fear, and material insecurity are palpable on nearly every page she wrote. For the family historian interested in the warts-and-all approach, such letters give more than enough details to tell a realistic tale of material hardship and domestic discord.

Minutiae and anecdotes abound in Brown's letters. There are long paragraphs about what kind of chickens Tam Scott preferred to raise, the contents of Marion's dreams and Aunt Agnes's late-night reminiscences, the family pig's name (Wallace), and which of the young Scott children played a grandmother (and wore green spectacles) in an 1899 children's pageant. We can know that a certain photo among many unlabeled family portraits is an image of Marion and her cousin Nellie Glencross, because an 1877 letter describes the sitting in sufficient detail—including the comment, "I think it is like me but they have made me squinting and I am sure I do not squint" (Brown to M. Glencross Bryden, 5 November 1877). For some family historians, no detail is too small to cherish, but these are the same everyday elements of family letters that have occasionally discouraged academic historians from considering the private correspondence of ordinary people as a worthwhile source.

Letters and the Historical Study of Immigration

Correspondence by its nature is a geographical process. Letters are mailed (or privately carried) from here to there, frequently recording a sense of distance, or longing for place. Some letters in archives contain long descriptions of places, complete with maps or references to maps (Gerber, 2006: 125-126; Richards, 2004a). A letter in the written version of a local dialect, or with spellings that reproduce regional pronunciations, also indicates a very specific place to the knowing reader. Other letters bear evidence of their origins and routing in their stationery, postage, postmarks, and enclosures. A long run of family letters, like those of Marion Brown and her relations, might trace the various residences and dispersal patterns of a large kinship group. Beyond such pinpoint data, the same run of family letters, perhaps especially when written by women (Dublin, 1981: 4-5), can narrate much about such geographical phenomena as the decision to emigrate (or not), and the experience of the transnational family.

A recent example of a study using family letters similar to the Marion Brown correspondence is by Greg Stott (Stott, 2006). Stott used private collections of nineteenth-century letters from a Canadian family, the Dunhams, to complicate the story of westward migration. Stott finds that an unmarried woman, a daughter caring for elderly parents, stayed in place, but held together a widely dispersed family through correspondence, for both practical and emotional reasons. Even when the contents of letters have little wider importance, the mere existence and survival of such a correspondence over decades is testimony to a sense of family that persisted despite distant settlement and other changes in circumstance.

Stott is hardly alone in his approach. In August 2003, the conference "Reading the Emigrant Letter," was held in Ottawa, under the sponsorship of the Carleton Center for the History of Migration. In Ottawa, scholars from around the world, and from a wide range of disciplines, gathered to share and discuss research on correspondence between emigrants and their homelands, correspondence among immigrants, and other similar topics. Among the conference presenters was David A. Gerber. Perhaps no historian has theorized the immigrant letter as text more completely than Gerber, in journal articles (1997, 2000, 2001, 2004) and a recent book (Gerber, 2006).

Gerber starts with the acknowledgment that immigrant letters are the "most widely proliferated and in volume the largest source we possess of the writing of ordinary people" (Gerber, 2000). But too often, he notes, such letters are only excerpted to "provide color and drama to historical narratives," without being considered seriously as a genre (Gerber, 1997). Instead, Gerber argues across several writings that there is an epistolary discourse in these letters that rewards a more critical reading. Gerber chooses a geographic metaphor to explain his idea, saying that long-term personal correspondence is

> a space mutually crafted over time by correspondents. Such letters break down conventional boundaries of place and create an alternative time dimension. They establish their own chronology of sending and receiving by which an epistolary relationship is charted, and through narration, they project the past into the present and the future. (Gerber, 2001)

In the matter too often dismissed (or even edited out, if the letters are published), Gerber finds negotiations over frequency and quality of contact, rituals of greeting and assurances of affection, simulations of intimacy, instructions reflecting levels of trust and expectations of privacy. What seems at first to be trivial business may in fact be the letters' most important content. "Personal problems in the human relationships of families," asserts Gerber, "may be the major undiscovered dimension of immigration history" (Gerber, 2000). Letters reveal these dynamics, and explain in some cases why some people became emigrants, and others, people much like their emigrant peers in every measurable sense, did not.

Or in the case of Marion Brown, letters capture a decades-long struggle with the very ideas of emigration and family. Her letters are not from an immigrant, but to immigrants; in David Gerber's formulation, that makes them a rare variety, but still part of the transnational conversation. "Homeland correspondents usually form an echo we faintly detect," he comments (Gerber, 2004), encouraging researchers to listen for that echo anyway, and consider the impact of these transatlantic exchanges on both sides. Brown's letters feature all the elements Gerber identifies as comprising epistolary ethical discourse: explaining delays ("I was nearly blind for five weeks," "I have had a burned hand"), negotiating conveyance, setting privacy expectations ("I don't want you to say a word about what I have told you"), reporting emotional states related to receiving or sending letters ("she sat up in bed and says to me read quick and let me know if they are all well"), excusing poor handwriting or materials, and contextualizing news (Brown letters dated 28 July 1869, 12 September 1872, 5 September 1881, and 13 December 1886, respectively). That her letters follow the script so well does not negate their power; rather, it gives the historian a context in which her writings can be interpreted with more insight.

What do Marion Brown's letters tell us about her experience of *not* emigrating? Brown's letters seldom dwell on this matter, so we must read them as Gerber instructs, considering the words and subjects she chooses with care. When she writes to her American cousins, Marion writes along a careful line, navigating between complaints of her troubles on one side and cheerful, helpful, news and counsel on the other. Many a sad report of hardship is followed by a defusing religious lesson, in which she indicates both resignation and reassurance:

> I took an ill turn and since that I have had no power in my left arm and I am lying in bed very helpless and sometimes i feel my place very hard but we must not grumble at God's dealing with us and must try to be content with whatever is to be my lot as long as I am in this world ... (Brown to "dear friends," 7 June 1880)

Especially when writing to her younger cousin Marion, she is eager to point out the moral lessons in her reports. Brown rarely reported church attendance (and at least one of the mentions of church attendance is disparaging), but she apparently claimed moral authority on the subject of hardship from her personal experiences, and the lives she witnessed at close hand.

Cousins Marion Brown and Marion Glencross never met. Marion Glencross's continuing correspondence was an emotional and material lifeline for Marion Brown, and the structure of her letters reveals that Miss Brown put a high priority on being a useful link to the American cousins as well. She could not, like some homeland correspondents did, provide landscape description or news of public events; Brown's everyday world was too restricted to allow this. She may never have seen a city as large as Dumfries, let alone Glasgow or the scenic parts of the Scottish countryside. She identified as Scottish in general, and imagined that that identity would possibly make a move to Dunmore easier: "Dunmore will scarcly be like a strange place now there is so many Scotch people in it", she projected (Brown to friend David Williamson, 1 December 1870).

For her young kinswoman, she made other efforts. Knowing that the American Marion was raised without a mother or older sister, the Scottish Marion took trouble to fill some of that feminine cultural void, by sending over dressmaking patterns, moral instruction, samples of local knitting, talk of dancing, instructions on removing creases from velvet, and advising John Glencross to teach his daughter the value of education:

Tell her to be anxious at the school for it is the best future you can give her to make her a good scholar for when once learned no one can take it from her and no saying what we are to need[;] if Uncle James had not made me a moderate good scholar what would have become of me now when I have to write everything I want to say... (Brown to J. Glencross, 23 December 1869)

There was a bit of self-interest in this advice, of course—an educated cousin would be more able to maintain the correspondence Brown hoped to continue (see Bell 2000, on Scottish emigrants and literacy in the nineteenth century). Before James Bryden of Ayrshire went to America in the 1872 to marry Marion Glencross in Dunmore, he visited with Marion Brown at Sanquhar, who found him more than suitable for her younger kinswoman (as a sudden turn to pretty stationery may attest): "I would kick up a right spree," she promises, if she could somehow attend their wedding (Brown to J. Bryden, 25 October 1874). In a transatlantic family, Marion Brown played the part of homeland correspondent and female mentor well.

But perhaps the strongest theme in the Marion Brown letters is a struggle over whether to emigrate, and lose that homeland correspondent status altogether. For much of Marion Brown's adult life, money was available for her passage, offered by Marion Glencross and her husband James Bryden; and she assured them it was a welcome offer, that she wanted to go to America badly, for their wedding, to hold their children and sing them lullabies ("bairns all like noise," she noted in a letter dated 8 January 1877), to eat apples from their tree and even polish their floors by dancing a jig. Opportunities to travel with others leaving Sanquhar were numerous. Nonetheless, Marion stayed in Sanquhar to the end of her days. Brown's ongoing health issues discouraged much hope: "I am not fit for the journey at this time," she explained in 1879, "for I am told there is 10 chances to one if I ever got to the other side of the water alive" (Brown to J. Glencross, 8 September 1879). Her fears were hardly unreasonable: women emigrants and those with disabilities did face significant physical risks and endangerment (O'Connell, 2000-2001; Richards, 2004b). Her sense of responsibility towards Agnes Scott added to the reluctance. To an offer of a chaperone, she answered, "I told him I could go to America with him, but for one thing. I could not leave Aunt" (Brown to M. Glencross Bryden, 10 June 1897). The chronological span of these comments tells the tale: her first surviving mention of the idea of crossing to America comes in an 1869 letter, when she says "I was telling Aunt if I could have walked any I would have gone to America with John White" (Brown to J. Glencross, 20 September 1869). Almost thirty years later, she confides to her cousin, "Many a time I wish Aunt had gone out to America when we left the Bogg as your Father wanted her but she would not" (Brown to M. Glencross Bryden, 20 October 1898). The last such comment to survive appears in 1902, where she tells Marion Glencross Bryden, "how I could like to see you face to face and have a talk with you" (Brown to M. Glencross Bryden, 12 January 1902). On a winter day, the isolated Scottish Marion, writing from a telephone office, yearned for a visit with the widowed American Marion, a woman she had never seen and knew only through letters and family lore.

Sometimes, this yearning took the form of poignant dreams, which Marion reported in her letters to America. To tell someone "I was dreaming about you last night" is a flattering confession of intimacy, in itself likely to fit Brown's bigger goal of maintaining family feeling. But the dream stories she tells are more specific—and in the following dreams about her cousin's fiance James Bryden, revealing:

> I dreamed the night before I got your letter that I got a registered letter and a nice ring in it and my name on the inside so you see I am stil going on with my dreaming yet. Do you ever talk any through your sleep now. I hope you will not tell them any queer stories about love at any rate. (Brown to J. Bryden, 10 October 1872)

> How is your Jeamie getting on I was dreaming about him the other night I thought he was come to take me away to America but when I wakened it was but a dream but I was quite disappointed when I wakened and did not see him... (Brown to M. Glencross, 31 October 1872)

> I was dreaming about you last night and was cleaning your boots and where do you think I was. I will tell you. I was in America but I saw no one I knew but yourself. (Brown to J. Bryden, undated, possibly 6 October 1873 by context)

From the safe distance of Sanquhar, Marion Brown told about dreams of love, boots, and an engraved ring from a young man about to marry her cousin, a young man who visited Marion at Sanquhar often before his emigration. Did the long distance (and Marion's poor health) render such flirtations harmless, even silly? Whether or not they seemed foolish, Brown wrote these dreams into her letters, often on pretty stationery, and thus left this evidence of her capacity for romance, and that romantic longing and disappointment were factors in her continuing correspondence with the Brydens.

Why did some emigrate, and some stay behind? It is a question historical geographers have asked, and tried to answer, for decades (Ostergren, 1988: 23). In the letters of Marion Brown, we get the kind of complicated explanation that ordinary life often brings, and correspondence records far better than census data or passenger rosters could capture. For Marion, there was no dramatic resolution, no crushing disappointment, no direct correlations. She wanted to emigrate, she had opportunities, but she also had obligations, and she was afraid for her own health. She did not go, but sometimes wished she had.

Marion Brown's last letter in the collection is addressed to her cousin Marion Glencross Bryden and dated 17 May 1903. In it, she begins with the emotional epistolary discourse David Gerber identifies:

> I felt a little surprised when I saw your welcome letter but I must say it was a pleasant surprise and had to hold it in my hand and look at it before I could muster courage to open it wondering all the while whither it would contain good or bad news.

She describes her health as "not very well", relating a day in bed waiting for a knee plaster to set. "Better to have a plaster knee than no knee at all", she jokes, to defuse the pathos of the scene. She reports on the doings of Tam Scott's ten living children, and asks after common acquaintances in Dunmore. And she thanks Marion Bryden for a monetary gift. No mention of Aunt, because Agnes Glencross Scott died in 1902, at the age of 85. Nothing unusual in this letter, no indication that it would be the last; perhaps it is only the last surviving letter. Marion Brown lived until October

1915, when she died and was buried in the Sanquhar Kirkyard, later to share a black marble tombstone with her cousin Thomas Glencross Scott (in 1928) and his wife Robina Boyle (in 1920), and two of the Scott sons (Hutchison, 2001).

Finding More Marion Browns

The collected Marion Brown correspondence is extraordinary. Not just for its contents, but for its mere existence. When David A. Gerber writes about the homeland correspondents of immigrant letter-writers, he writes "we may at times hear the echoes" of their stories in the immigrants' return letters, but "such double-voicing is relatively rare" (Gerber, 2006: 7), There is no need to read through immigrant letters to find Marion's story; we have over a hundred letters in her own hand, carrying her own voice. Gerber also points out that most of the letters written by British emigrants were written by men, a finding also seen in Charlotte Erickson's earlier work (Erickson, 1972: 241). They argue that in the nineteenth century, women in Great Britain would not have had access to as much schooling as their male peers, so that "women might have been readers, and they might have learned to sign their names, but they may have lacked access to learning the skills for doing extensive writing," speculates Gerber (2006: 82). In addition to the problem of production, there is some reason to believe that women's private writings are less frequently preserved (see Kimbell Relph, 1979: 598; Domosh and Morin, 2003: 262; Beattie, 1997). Marion Brown probably had little formal schooling, but she clearly had a "nice hand" for correspondence; it is possible that her inability to speak for long periods made writing a more important skill for her to develop. Collections like those studied by Greg Stott are considered strong at 64 letters, but the Marion Brown correspondence includes well over 200 letters from Scotland and Scottish immigrants (more than half by Brown herself).

So we have a rare collection under study here. But must it remain quite so rare? One goal both family historians and historians of immigration can agree upon is the need to find more collections of letters like Marion Brown's, and preserve them for future research. "We need to bring emigrant letters out to study them more often," notes Bill Jones, "those large and laden letters ...still await and deserve our responses" (Jones 2005: 46). Among archival sources, the letters sent *to* American immigrants are, so far, vastly outnumbered by those sent *by* American immigrants to their homeland friends and family. But this imbalance does not mean such letters do not exist anywhere. The Marion Brown letters are still in private hands, as are the Dunham letters consulted by Greg Stott (2006). How many other such treasures remain to surface, or disintegrate, in hot attics, damp basements, and boxes too easily discarded? There is much to be gained by a more organized effort to collect such materials, and publicize their value. "Consciousness of the importance of historical documentation does not come automatically," noted Margaret Strobel, in a description of a project at the University of Illinois at Chicago, called "Don't Throw it Away!!" (Strobel, 1999). While Strobel's project targeted the archives of activist organizations for preservation, it may offer a model for a similar appeal for family archives.

Genealogical organizations and state historical societies are among those proving instructions and encouragement to families with old papers to preserve. The Minnesota Historical Society, for example, has a Conservation Outreach program, which conducts public workshops on protecting family history materials. Its website on "Preserving your Family Letters and Paper Heirlooms" includes such crucial information as optimal storage conditions, safer display and storage products, and sound advice such as "Handle your important paper items by their edges and with clean hands to avoid soiling the surface" (Minnesota Historical Society, 2006: np). The national all-volunteer Legacy Project seeks to teach Americans to preserve wartime correspondence. The "How to Preserve Your Letters" page on its website assures owners of such letters that they are indeed valuable historical documents, even if their contents are very personal; but, it continues, "many of these correspondences are being thrown away, lost, or irreparably damaged" (The Legacy Project, 2005). Their preservation instructions remind private collectors not to laminate, staple, glue, or rubberband original artifacts, and suggests that owners to save envelopes—and make hardcopies of emails, from more recent conflicts. Such basic, detailed, online instructions have the potential to reach a vast unaddressed need.

Of course preservation is the first priority—without the documents, there can be no further exploration of the correspondence—but a well-kept set of letters will still be of little use to researchers unless its existence is somehow made known, and access to the letters in some form is granted by their owners. Some owners can be persuaded to donate a collection to a state historical library, or hand it over to archivists in a university special collections department. Others are willing, even eager, to transcribe and publish the contents of letters, or even post the transcripts online, although the skill of the transcriber may vary, and verification remains a challenge. Beyond those avenues, some family-held collections will always remain private collections, for many reasons, but even those may be made available for study purposes upon inquiry, on a case-by-case basis. "The best local documents," noted historian Raphael Samuel, "will often be found not in the library or the record office, but in the home" (Samuel, 1976). As with most research, asking the right people the right questions will often bring a wealth of surprising sources to light.

Not only is someone like Marion Brown of interest to family historians, she *was* in a sense a family historian, or even a family geographer, making and marking a personalized space between homeland and abroad. Lorimer (2003) writes about grass-roots producers of geographic knowledge, and on the need for recovering evidence of a relational understanding of place. Marion's letters preserved family history details from her own household, such as birthdates, causes of death, and dates of marriage. She recorded messages dictated by her aunt, Agnes Scott, who was probably not herself literate, and thus preserved a key older family member's words and memories. She provided photographs of some of the people she named, and collected information from America for sharing with branches of the family in Scotland. She wrote about the distance between herself and her correspondents, and how that distance felt. Marion Brown illustrates a common function of single women in large families—she serves as conduit, recorder, and disseminator of the family's story as it unfolds.

Marion Brown was knitting a family history along with her stockings and sweater vests. Her letters to Marion Glencross Bryden were passed along to her daughter Helen Bryden Marsh (1887-1968), and then to Helen's daughter Louise Marsh Richards (b. 1919), and then to Louise's granddaughter, the present author, Penny Richards (b. 1966). This story is a women's story and a family story, from the moment a pen touched paper in Sanquhar, until the moment fingers tapped a keyboard in California to write these words. The written threads worked by Marion Brown are indeed her most lasting legacy.

Note: The Marion Brown letters remain in the author's possession—in acid-free folders, in an acid-free archival box, on a shelf in a cool, dark closet. They are rarely handled or exposed to light because they were transcribed in 1994, by the author.

References

Bartholemew, J.G. (1895), *The Royal Scottish Geographical Society's Atlas of Scotland*, Edinburgh: Geographical Institute.

Beattie, D. (1997), "Retrieving the Irretrievable: Providing Access to 'Hidden Groups' in Archives", in Cohen, L.B. (ed.), *Reference Services for Archives and Manuscripts*, New York: Haworth Press, pp. 83-94.

Bell, B. (2000), "Crusoe's Books: The Scottish Emigrant Reader in the Nineteenth Century", in Bell, B., Bennett, P. and Bevan, J. (eds), *Across Boundaries: The Book in Culture and Commerce*, Newcastle, DE: Oak Knoll Press, pp. 116-129.

Bredberg, E. (1999), "Writing Disability History: Problems, Perspectives, and Sources", *Disability and Society*, 14(2): 189-201.

Breitenbach, E., Brown, A. and Myers, F. (1998), "Understanding Women in Scotland", *Feminist Review*, 58: 44-65.

Brown, J. (1891), *The History of Sanquhar*, Dumfries: J. Anderson and Son.

Cassidy, F. (ed.) (1985), *The Dictionary of American Regional English, Volume I: A-C*, Cambridge, MA: Harvard University Press.

Domosh, M. and Morin, K. (2003), "Travels with Feminist Historical Geography", *Gender, Place, and Culture* 10(3): 257-264.

Dublin, T. (ed.) (1981), "*Farm to Factory: Women's Letters 1830-1860*, New York: Columbia University Press.

Dumfries and Galloway Museums Service (1998), "A History of the Sanquhar Knitting Pattern", http://www.dumfriesmuseum.demon.com.uk/knithist.html (accessed October 18, 2006)

Erickson, C. (1972), *Invisible Immigrants: The Adaptation of English and Scottish Immigrants in Nineteenth-Century America*, Coral Gables, FL: University of Miami Press.

Gerber, D.A. (2006), *Authors of their Lives: The Personal Correspondence of British Immigrants to North America in the Nineteenth Century*, New York: New York University Press.

Gerber, D.A. (2004), "What is it We Seek to Find in First-Person Documents? Documenting Society and Cultural Practice in Irish Immigrant Writings", *Reviews in American History*, 32(3): 305-316.

Gerber, D.A. (2001), "Forming a Transnational Narrative: New Perspectives on European Migrations to the United States", *The History Teacher*, 35(1): 61-78.

Gerber, D.A. (2000), "Epistolary Ethics: Personal Correspondence and the Culture of Emigration in the Nineteenth Century", *Journal of American Ethnic History*, 19(4): 3-23.

Gerber, D.A. (1997), "The Immigrant Letter between Positivism and Populism: The Uses of Immigrant Personal Correspondence in Twentieth-Century American Scholarship", *Journal of American Ethnic History*, 16(4): 3-35.

Hutchison, I. (2003), "Disability in Nineteenth-Century Scotland: The Case of Marion Brown", *University of Sussex Journal of Contemporary History* 5 (January): available online at http://www.sussex.ac.uk/history/documents/iainhutchison.pdf (accessed 25 October 2006).

Hutchison, I. (2001), Email correspondence and photographs sent to the author, pertaining to the Sanquhar Kirkyard and nearby cemeteries.

Jolly, E. (2005), "Grandfather's Memorabilia: A Fragment of Victoria's Maritime History", *The Genealogist*, March: 172.

Jones, B. (2005), "Writing Back: Welsh Emigrants and their Correspondence in the Nineteenth Century", *North American Journal of Welsh Studies*, 5(1): 23-46.

Kimbell Relph, A. (1979), "The World Center for Women's Archives, 1935-1940", *Signs: Journal of Women in Culture and Society*, 4(3): 597-603.

Legacy Project (2005), "Preserving your Wartime Letters", http://www.warletters.com/preserve/index.html (accessed 25 October 2006)

Leneman, L. (1999), "The Scottish Case that Led to Hardwicke's Marriage Act", *Law and History Review*, 17(1): 161-169.

Lorimer, H. (2003), "Telling Small Stories: Spaces of Knowledge and the Practice of Geography", *Transactions of the Institute of British Geographers*, 28(2): 197-217.

Minnesota Historical Society (2006), "Preserving your Family Letters and Paper Heirlooms", http://www.mnhs.org/people/mngg/stories/papers.htm (accessed 25 October 2006).

O'Connell, A. (2000-2001), "'Take Care of the Immigrant Girls': The Migration Process of Late-Nineteenth-Century Irish Women," *Eire-Ireland*, 35(3/4): 102-133.

Ostergren, R. (1988), *A Community Transplanted: The Trans-Atlantic Experience of a Swedish Immigrant Settlement in the Upper Middle West, 1835-1915*, Uppsala, Sweden: Acta Universitatis Upsaliensis.

Richards, P.L. (2004a), "'Could I But Mark out my Own Map of Life': Educated Women Embracing Cartography in the Nineteenth-Century American South", *Cartographica*, 39(4): 1-17.

Richards, P.L. (2004b), "Points of Entry: Disability and the Historical Geography of Immigration", *Disability Studies Quarterly*, 24(3), online at: http://www.dsq-sds.org/2004_summer_toc.html (accessed 25 October 2006).

Samuel, R. (1976), "Local History and Oral History", *History Workshop Journal*, 1(1): 191-208.

Schulze, L.M. (2002), "Past Voices: Letters Home", a section of the Olive Tree Genealogy website, http://www.pastvoices.com (accessed October 18, 2006).

Stott, G. (2006), "The Persistence of Family: A Study of a Nineteenth-Century Canadian Family and Their Correspondence", *Journal of Family History*, 31(2): 190-207.

Strobel, M. (1999), "Becoming a Historian, Being an Archivist, and Thinking Archivally: Documents and Memory as Sources", *Journal of Women's History*, 11(1): 181-192.

Thomson, A. (nd), "Sanquhar's Traditional Knitting", (a pamphlet provided to the author by Iain Hutchison).

Young, C. (1991), "Women's Work, Family, and the Rural Trades in Nineteenth-Century Scotland", *Review of Scottish Culture*, 7: 53-60.

Chapter 6

Remaking Time and Space: The Internet, Digital Archives and Genealogy

Kevin Meethan

Introduction: Recovering Kinship

The development of information and communications technology over the past two decades, in particular the spread of the Internet, has brought with it many changes. To Castells (2000: 3) it is no less than a global revolution that has reshaped the economic, social and cultural realms in an unprecedented way. "In a world of global flows" he adds, "...the search for identity, collective or individual, ascribed or constructed, becomes the fundamental source of social meaning". Castells implies that the old certainties of place and territory that equated cultures with national or sub-national boundaries, of belonging to a nation, are under threat or are being replaced by newer, more individualised forms of identity that cut across the established categories of culture and place. While that may be the case—in some circumstances, such generalisations also need to be treated with caution. Katz and Rice (2002: 6-12), for example, note three social issues that are central to Internet use: access, civic and community involvement, and social interaction; which are often surrounded by inflated rhetorics of both a dystopian and utopian kind that seemingly allow little room for compromise. Kitchin (1998: 16) voices similar concerns, and notes that while the Internet has led to many fundamental changes, like other developments subsumed under the term "globalisation" these changes are neither uniform in their effects nor evenly distributed. While new communication technology recasts the relationship between people and locality by shrinking time and space, rather than undermining the importance of locality as a source for identity, it may instead reinforce it (see also Dicken, 2003; Kraidy, 2005; Meethan, 2006).

New technological, social and cultural changes inevitably bring both positive and negative aspects. Champions of the positive benefits point to the opportunities new technologies offer for overcoming the constraints of place and time, and the ways in which networks and "communities" can exist on the basis of shared interest regardless of spatial constraints (Foster, 1997; Green et al., 2005; Mitra, 1997). As with any other form of social interaction, the Internet also "...enables people to use cultural attributes to recognize themselves and construct meaning" (Katz and Rice, 2002: 13). As Green et al. (2005) point out, one of the defining characteristics of the Internet is that it can transform space by helping to overcome the limitations of location. An example of specific relevance here is Miller and Slater's (2000: 58-60) ethnography of Internet use in Trinidad. They found that the Internet enabled long-term relationships between distant kin, both genealogically as much as spatially, to be viable in new ways. As Robins and Webster (1999) argue, the use of the Internet has led to a new form of "knowledge space", which is fluid, open and dynamic, and also allows direct and immediate contact. They also caution that such forms of knowledge space do not replace localised forms of knowledge, but rather exist alongside them.

This chapter examines Internet phenomena in relation to the growing leisure pursuits of family history and genealogy. Although family history is an increasingly popular pastime, it is difficult to gauge the exact number of people in active pursuit of their lost ancestors, and few academics are concerned with people who pursue family history. Notable examples include articles by Drake (2001), Jacobson (1986), Jacobson et al. (1989), Lambert (1996, 2002), Nash (2002) and Meethan (2004). Quotations from family historians presented in this chapter have been extracted from an online survey that attempted to define the general characteristics of online family historians, and a subsequent email dialogue with selected informants of European descent, who had travelled to Europe to rediscover, or recover, their family histories (as described in Meethan, 2004).

Within family history, approaches range from researching one family name to researching the ancestry of multiple maternal and paternal lines. Relationships can also be traced horizontally—across families, which can often result in the discovery of living relatives, as well as vertically—that is, back in time. Whichever approach, or mixture of approaches, is adopted, the overall aim of family history is a search for the authentic and verifiable material trace of lineages that connect the individual in the here and now to other people in other times. Using much the same methods and resources that professional historians employ, family history relies on an extensive array of national public records and other archival sources that contain the evidence that links the present to the past. Like other forms of history, family history involves gathering and collating the material traces of artefacts and documents such as birth, marriage and death certificates, parish and other church records, immigration records and census data, which demonstrably link people together, establishing proven lineages of descent (Herber, 2005; Hey, 2004; see also http://www.nationalarchives.gov.uk/).

While the development of microfilm and microfiche from the 1920s onwards enabled easier storage and retrieval capacity for some archival sources, accessing and using the data required the time and the means either to travel to the archives, or to employ someone else to undertake the work. One of the most valuable archives in terms of its general extent and availability, is provided by the Church of Jesus Christ of Latter-day Saints, or the Mormon or LDS Church (see http://www.familysearch.org/ and Otterstrom, this volume), for reasons connected with its beliefs. It has amassed a huge archive of vital and historical records in Salt Lake City. Microfilms and microfiche of many of these records can be viewed at one of the LDS local Family History Centres (FHC) that exist across the world. Members of the public may also visit a local FHC and use its facilities without cost, although voluntary donations are welcomed. Films and fiche not held locally can be ordered from the main archive in Utah at a small cost, cannot be removed from the premises and remain the property of the local FHC, so that each FHC can build up its own archive.

The more recent digitisation of archive material and its easy availability through the Internet has made a significant impact on family history, not only creating its own markets and economies (see below) but also contributing to a more fundamental reshaping of the ways in which individuals can retrieve information and create new ways of relating to history and to place (Castells, 2000; Mitra, 1997; Foster, 1997; Katz and Rice, 2002; Kitchin, 1998). Examples include the UK Civil Registration Index for births, marriages and deaths up to 1919 (http://freebmd.rootsweb.com/). Other sources such as newspapers, trade and street directories, ships' passenger lists and other sources are rapidly being added (see http://www.rootsweb.com/), while older sources such as the contents of the Doomsday or Domesday book, compiled on the orders of William the Conqueror in 1085, and which lists all the landholdings and landowners in England at that time, can now be accessed and searched online (http://www.domesdaybook.co.uk/index.html). While some of this activity is being undertaken by national, regional and local government agencies, much of it is also carried out by dedicated volunteers who undertake the laborious activity of data entry from paper and micofiche/microfilm records to databases, while the easy availability of web space also means that information relating to individual family trees is often posted on websites and in the public realm. Of course barriers remain, most notably those of language and access to the Internet itself, yet today it is entirely possible for people to undertake family history research without actually visiting or consulting the original archives. Misztal (2003: 47) terms one of the consequences of such developments a "…process of denationalisation of memory as well as trends towards the fragmentation and democratisation of memory".

Modernity, Bureaucracy and the Archive

Family history is above all a form of memory work (Lambert, 2002), a rediscovery of lost ties, a means to identify, catalogue and arrange the unknowns of one's personal past (Halbwachs, 1992) as clearly there is a need to undertake family history research only if connections to the past have been lost. Before the advent of the modern and comprehensive state bureaucracy in Europe, surviving connections to past generations were usually provided by immediate family and ties of kinship. They were, by and large, transmitted orally and were strongly linked to a specific place of residence and a sense of territorial identity for families who remained relatively stationary over time. In turn, these links were cross-cut by other forms of stratification such as gender roles, social class and religious affiliation. The apparent anchors that these distinctions created provided for better or worse a matrix of identity and belonging for many families that modernity appears to have lessened or even swept aside (Beck, 1992, 1994, 2000; Giddens, 1971, 1991, 1994) in places where tradition is disappearing, and forms of perception become deracinated and ahistorical (Giddens, 1994; Halbwachs, 1992; Hoskins, 2004; Huyssen, 1995; Nora, 1992).

The reckoning of descent can also be subject to both social and political pressures (Eriksen, 1993) and also shows considerable variation both within and across cultures (see Adams, 2006: 43-53, and Kuper, 1999). Even allowing for outright fraud and the possibility that lineages can be falsified and doctored for certain ends, in the past the accurate recording of lineages in documentary form in European societies was the preserve of aristocrats, whose desire to hold on to privileged positions, status, money and property, required proof of pedigree or "breeding" to maintain the lawful transmission of status and patrimonial property rights. As Connerton (1989: 86) notes, "…the value of blood is that of a sign", and what it signified, and still signifies to some extent, is social standing. Such status, however, required authentication and proof, which in turn required accurate written records. Indeed the recording of an individual was often a way for elites to assert control if not ownership of other human beings. Czap (1983) notes that the Russian "soul revisions" undertaken between 1780 and 1858 not only enumerated the population for purposes of taxation, but also located individuals within a strict social hierarchy based on serfdom, while slaves in the Caribbean and USA were regarded as property and recorded as such in wills and probate records.

Archives of course come in many types. Some comprise the paperwork and records of famous individuals; some detail the activities of religious communities, companies and other significant private and state organisations; whereas others contain collections of newspapers, photographs and other media. Geographical scales of archival coverage also range from the national to the local. All these sources have potential significance for family historians. Notably, however, the modern bureaucratic state, with its emphasis on census enumerations or the need to record and archive accurate birth, marriage, and death records, now assists people in such countries, or whose ancestors lived in those countries, to trace their families back at least a few generations where other non-governmental genealogical data sources such as parish registers are lacking (Dandecker, 1990; Diamond, 1999; Kraeger, 1997; Woods, 1996).

The establishment of national records on ordinary citizens went hand in hand with the origins of the modern nation state. For example, national registration of births, marriages and deaths was introduced in France in 1792 (http://www.geneaguide.com/anglais/catholic.htm), and the first census of the USA was undertaken in 1790. Initially the U.S. Constitution mandated a census every decade as a means of determining each state's number of members in the federal House of Representatives. As Kraeger (1997: 155-6) notes, however such developments subsequently provided a source of demographic data to inform policies, and were also a means by which the correspondence between a demarcated territory and a defined population as the nation could be demonstrated. To be recorded and enumerated was to belong to a particular place.

During the course of the nineteenth century and as a consequence of urbanisation the established patterns of work, kinship, home, family and locality that had previously dominated social interaction underwent a fundamental transformation in industrializing countries. The rhythms of agricultural time and sense of deep and long connection to a rural locality gave way, for those enticed to the emerging cities, to the dominance of regulated clock time which controlled the new centres of urban production (Sussman, 1999). In addition there was a widespread overseas population movement, most notably from Europe to North America and to colonies in Africa and Asia.

Parallel to these changes was the development of state bureaucracies, regulatory systems that required the accurate ordering and gathering of data at both a personal and national level, which in turn involved the development and increasing importance of a number of mediating institutions that regulated economic and social activity, for example the World Bank, the U.S. federal government, and the United Nations (Polanyi, 2001). The act of recording also offers the capability, via documents, to transcend the limitations of local space and time (Le Goff, 1992: 58-60) and thereby to construct larger frames of reference such as nations, states and peoples through the spread of mass literacy and civics education. For example Jedlowski (2001: 37) notes how these developments contributed to the "exteriorization" of memory; memory was no longer only that which was held, so to speak, only in an individual's recollection of events, but rather that which was written down and authorised, a process further accelerated by the print media.

Oral memory nevertheless did not entirely succumb to what has been termed documentary memory (Bal, 1999; Freeman, 1993; Le Goff, 1992). First, although not all cultures or families rely on archive sources for the tracing of ancestry; indeed it is precisely because more traditional types of links and forms of reckoning descent have been lost that family historians today resort to the archive as a means of recovery. Second, recent research into the use of autobiography and memory in identity formation shows that the negotiation of identities through oral transmission, and the interplay between written and oral sources is complex and dynamic, shaped in part by the social, cultural and political circumstances in which it is located (Basu, 2004; Chamberlain and Leydesdorff, 2004; Favart-Jordan, 2002; Pálsson, 2002; Tyler, 2005; Wang and Brockmeier, 2002). For those living in the developed economies of modernity, the further back in time one pursues family history, the greater the importance of archives, whether belonging to government, church, or volunteer organizations, in preserving the material trace of one's ancestry.

National archives are a consequence of a bureaucratic system as well as a repository of the raw materials out of which a sense of nationhood, national myths and narratives can be forged. In this sense, one can talk of "public" or "collective" memory as forms of history and discourses that emphasise solidarity within a larger unit strangers with whom one nonetheless feels some affinity in the way that Anderson (1991: 5) described as an "imagined community":[1]

> It is *imagined* because the members of even the smallest nation will never know most of their fellow-members, meet them, or even hear of them, yet in the minds of each lives the image of their communion. (emphasis in original)

Anderson also draws attention to the ways in which a sense of community is achieved through narratives and discourses of the nation and history[2] that bind people into forms of collective identity. He attributes it in part to the rise of "print capitalism", that is, the popular press in addition to other forms of non-oral communication that enabled geographically dispersed populations to share a common sense of nationhood. The state as nation becomes the spatial referent for a wider sense of solidarity and belonging. This process is part of the long-term shift that over the centuries tends to

---

1 Anderson also states that "imagined" in this sense should not be taken to imply that they are therefore "false" or inauthentic.
2 For forms of diasporic identity, see Coles and Timothy (2004).

replace oral memory with documentary memory. Nora (1996) sees this "acceleration" of history through modernity as encouraging forms of collective memory to become mediated, or indeed only possible, through the existence of documentary archives and other types of media (see also Hoskins, 2004; Thompson, 1995). Oral memory, or memory as the collective lived experiences of people, becomes fragmented and recoverable principally through the traces in the archives.

Today the creation of a sense of shared history and solidarity that relies on the past also relies on the capacity to store, catalogue and retrieve information. The existence of an archive becomes a powerful adjunct for a sense of collective identity as history that cuts across localised forms of kinship, memory and belonging. Once that process has been established, a national archive becomes a means by which it is maintained, by providing the legitimacy of the state with origins, precedents, rights and obligations as enshrined in documents. Of course there are many other "archives" besides national archives, including local libraries, historical societies, old city directories, churches, and even the living memory of elders.

In common with church-based and other local depositories, a national archive, as a repository of knowledge around which citizens learn their sense of collective identities, is neither neutral nor all encompassing, for to be admitted, the documents that comprise the archive have to be prejudged as worthy of inclusion, as only a fraction of those produced will survive. "The document" as Le Gof (1992: xviii) writes, "…is not objective, innocent raw material, but expresses past society's power over memory and the future: the document is what remains", and it is also the document that provides the visible, tangible and proven link between the present and forgotten ancestors of the past (see also Bradley, 1999; Caygill, 1999; Osborne, 1999).

Although there are similarities between family history and professionally-written history, in that they both rely on archives and other forms of material traces, family history is also a personalisation of history. The links to the past that are uncovered or recovered are as unique as genetic traits, being shared only with one's relatives or even only with one's siblings. Lambert (1996) terms this dual nature of family history as the distinction between historical time which is impersonal and "other" and exists in and through the archive, and autobiographical time, which consists of remembered experiences. The task of the genealogist, he adds, "…is to make autobiographical time continuous with historical time" (Lambert, 1996: 135). This is not usually the history of unfolding grand narratives (unless one's ancestors participated in major historic events) or of competing ideologies, rather it is the history of the mundane. Through the imprint in the archive that births, marriages and deaths leave behind—a trace however small links the individual in the here and now with a unique autobiographical past. As such the search for ancestors is both a process of recovery and discovery, a search for a form of "deep" kinship that extends beyond living memory and that is often lost in modern developed societies.

Autobiography, Biography and Memory

Memory is both a collective phenomenon that binds people together through forms of a social imaginary, and the unique legacy of each individual's unique biography.

For Halbwachs (1992: 52), individual memory was an aspect of collective memory, itself a "totality of thoughts common to a group". Misztal (2003b: 5) notes that Halbwach's formulation, with its Durkheimian emphasis on the group, sees collective memory as a form of deterministic moral solidarity and consensus. She reminds us, that societies do not remember (see also Bell, 2003), for after all, they are not endowed with agency. Memories are always situated in a social context, but this should not be seen as a simple or straightforward projection of a collective, shared memory. Rather the themes, metaphors and imagery are derived from commonly held written and visual sources act as a repertoire of shared forms that people draw on, providing the reference points around which individual experiences of the self can be made intelligible to others who share a similar background.

Biographical memory is the key to a sense of personal identity. Memories are accumulated and ordered into a narrative of the self, a continuous process of making and remaking the self, which as Cavarrero (2000) reminds, is always "work in progress". Similarly Hearn (2002: 748) writes that narrative is "...crucial for orienting and guiding behaviour, making both practical and moral sense of reality".

Narrative is the organising framework around which the stories of lives are woven, the ordering of memory into a coherent and explicable patterns. Cattell and Climo (2002) also point to an inherent tension between memory as the unique life story of an individual, and the individual as a "representative" of wider sociocultural formations. The production of a biographical narrative relies on the prior existence of forms of collective memory, which "...perceived as being continuous with the past provide a sense of history and connection, a sense of personal and group identities" (Cattell and Climo, 2002: 27). Memory, even of the most personal and embodied kind, is consequently at once both intersubjectively constituted and also situated within a wider and constantly changing frame of reference (Cavarrero, 2000; Chamberlain and Leydersdorff, 2004; Favart-Jordan, 2002; Jedlowski, 2001; Skultans, 1998).

Family history adds both a widening and deepening of the overall frame of reference or repertoire from which an individual can draw. The naming and identifying of past ancestors in time and place enables a wider contextualisation and linkages to be established between the here and now of the researcher, and the other places and other times through which collective memory is realised (Connerton, 1989). Of course no archive is entirely complete, and where gaps exist, family history researchers often use the wider context to "fix" their ancestors in time and place. For example, this can be clearly illustrated by the following comments of two of my informants, Mary and Ellen,[3] both of whom were from the USA (see Meethan 2004). Mary stated that

> My reading of Irish history, for me, provides a context for understanding the lives of the people whose names and dates turn up on my tree. In some cases, I have been lucky enough to obtain some old letters, possessions, or records to provide some documentation for their lives. But where no such records exist, history fills in a lot of blanks.

Ellen explained that

---

3   All informant's names have been changed to ensure anonymity.

...I started reading history and found out all kinds of stuff about the social conditions in England at that time [mid-19th century] which was a bit of a shock. Then I knew why you can find whole families in the death indexes that had died more or less at the same time.

These brief accounts also draw attention to the limits of living memory itself, for while the names of ancestors and perhaps their deeds may be known to their descendants, either through oral transmission or archived materials, active living family memory does not often extend back further than four or five generations at best. "The dead", as Halbwachs remarks (1992: 73), "...retreat into the past" and as such, their memory becomes obliterated, so that family memory becomes a constantly receding horizon.

Yet family history is as much about utilising the methods of history and establishing provenance and accuracy as is academic history. Barbara told me that

As I went along I found that the stories gramps had told me weren't quite true, perhaps his memory wasn't what it was but names and places and dates were all muddled up and led me down a number of blind alleys until I found the proof, none of it was written down you see until I started.

Stephen made much the same point:

My motivation for doing this (before I got hooked) was because many of the family stories seemed to be horse feathers. I was given a family history about ten years ago that was compiled by a distant cousin which confirmed some of the things I thought were horse feathers.

The family historian often wishes both to account for the material traces that can be located within archives and to set them within the wider narratives and discourse of nation and place, as well as to confirm the more personalised and idiosyncratic memories of family and kin through which elements of collective and individual memory are mediated. Such an undertaking means not only discovering linkages, but also balancing the archival traces of institutions with the personal narratives of self and other kin, using the same epistemological underpinnings that allow an objective and verifiable model of events to be constructed from the archival remains of the past. Another important aspect to family history is that in order to claim one's research as verifiable and objective, it has to exist in a material form. Archives proliferate. Further consequences of new technologies are that they enhance access to officially sanctioned records of state institutions as well as other sources (for example, parish registers and LDS Church records) at the same time that they enable the creation, storage and dissemination of countless personal archives, as for which family historians become the "the keeper of the memory" (Kellerhalls et al., 2002: 220, see also Bradley, 1999). In this sense state-approved memory, in conjunction with other locally and religiously-archived material, has not only, to use Jedlowski's (2001) term, become exteriorised, both in terms of sources and the way in which these are treated, it has also become more demotic and personalised.[4]

---

4   A more recent development in FHR is the use of genetic testing. See Pálsson (2002) and http://www.familytreedna.com/) and Chapter 9, this volume.

The Digital Archive

The linkages and sense of continuity with place that have been "undone" to some extent by modernity are in some cases being replaced and rediscovered through the undertaking of family history research. Ironically one of the means by which this is achieved is itself a consequence of modernity and the development of bureaucracy. The recent rapid rise of family history as a more mass and democratised pursuit is an unanticipated effect of the recent development of the Internet, home computing, and the subsequent availability of accessible archives in digital format.

Online search engines now provide access to material that was previously available only to dedicated of researchers with both time and money. The growing interest in family history must in part be due to the relative ease that ancestors can now be traced via the Internet, which in turn, has resulted in the creation of a large number of public agency, commercial, volunteer organization, and individual family historians' web sites related to genealogy. Commercial companies typically charge a subscription fee (e.g. http://www.ancestry.com) for their data, and their main archives are accessed via the Internet, although some records may also be purchased on CD. In addition, the Mormon Church-sponsored website (http://www.familysearch.org) allows the surfer freely to access indexes and locate specific records and sources.

Other accessible databases can also be located through Rootsweb, a voluntary organisation dedicated to the free dissemination of genealogical resources, and which hosts over 29,600 lists dedicated to genealogical research.[5] In addition, large numbers of bulletin boards and lists enable people with specific interests to exchange information and ask for advice. Most sites also offer tips and advice for those starting their research, and advise on protocol or "netiquette".

This growth of interest in genealogy has also resulted in the development of specific software packages that utilise the capacity of the standard PC to hold and order large amounts of data. The compilation of a large family tree can now be accomplished in a matter of minutes, and some packages allow for the combination of text with scanned photographs. As well as manuscript census records, other material such as town directories, immigration records, court and probate records and old newspapers are also being added to the global digital archive so that the tracing of one's ancestry is no longer so delimited by spatial or temporal constraints of site visits to archives.

Family historians, however, now use the Internet for more than just locating documents. New forms of "knowledge space" are emerging that are open, fluid and dynamic. Mailing lists and other forms of information exchange are freely available and in the public domain, which greatly assist the pursuit of family history, as Janet remarked:

> I guess I probably would have been working on my family genealogy; but it would have been a hit & miss, or a now & then thing before the computer explosion. Plus now that I am retired I have the free time ...the computer really opened it up after I found about "mailing lists" & all the help & advice available from the people on them.

---

5 Members may subscribe to multiple lists, but many are effectively moribund, having few members and very infrequent postings.

This account illustrates the importance of networking. Email lists are used either to request or share information, or to seek advice on the location of archives or other forms of information not so easily available to particular individuals. Emails posted on message boards may take the forms of a simple request such as "...does anyone know about X who lived in Y in the 1890s?" On occasion, people will circulate lists of names, places and dates that they have uncovered in the hope that they may be of use to other researchers, as another form of connectivity.

One of the claims for the use of the Internet in family history research is the ways in which it allows new connections to be made. Family historians always face the possibility that others also researching a particular name may be related to them: "... doing it got me back in touch with cousins I had not seen for years", as one informant noted; although informants who reported this kind of connection considered it a bonus rather than an end in itself. Undertaking family history can also engender a spirit of cooperation among amateur genealogists, especially when confronting the "brick wall", the point at which no progress is being made, as Susan explained:

> There are still brick walls to break & with the friends that I am now making through the net & the rellies [relatives] that I have found, I feel that this will be easier for me as I research my own roots & those of my husband.

The development of information technologies has created the means by which people can now reclaim their lost kin, and by so doing provide a tangible and proven connection between the immediate here and now and the more abstract temporal scale of history and collective memory. The Internet has also enabled the proliferation of personal and family archives, for new technologies not only provide the means of data recovery but also the means of preservation and transmission. As Caygill (1999: 2) points out, the development of a global archive, or to be more precise, the development of a series of archives which have global spread, raises the intriguing possibility that they will act as more than an efficient technique of recall, and instead "...mark the beginnings of a new, inventive relationship to knowledge, a relationship that is dissolving the hierarchy associated with the [national] archive".

Conclusions

The introduction to this chapter included a number of general issues and claims for the effects of information technologies within the context of globalisation. I also noted the need to treat with caution the more exaggerated and rhetorical claims associated with these developments. In the case of family history materials, the use of new technologies represents a very different way of imagining, in Anderson's (1991) terms, the relationship between the present and the past that cuts across the constraints imposed by space and time. By doing so, it helps to reconcile the place of the individual in the here and now with the collective memories that are stored and legitimised in the material traces that can be found in the archives.

While the digitisation of archive sources is far from comprehensive, a new form of Internet capitalism as represented by on-line genealogical services, analogous to the print capitalism proposed by Anderson (1991), is beginning to emerge. In turn, it provides the social imaginary with the tools by which new forms of solidarity and imagined collectivity can be created. The rediscovery and reclamation of history and collective memory through the Internet means that self identity is increasingly organised in a reflexive and negotiated fashion that may span many different places, times and nations (Giddens, 1991). Historical memory and the tools by which it is forged, become democratised and opened up across space and time. These new technologies do not result in a loss of collective memory, but in a reconfiguration. As modernity "destroys", it also provides the means to reinvent, re-imagine and recover ties of kinship and the past, in new ways. Global connectivity does not undo history nor "detraditionalise" the world, leading to a "brave new world" of cyberspace. Instead, the development of the digital archive offers ordinary people the means and capacity to imagine the past, to reconnect and reclaim history as their own, as unique and irreplicable in ways that their forebears never could have imagined.

While the Internet holds out the promise of more decentralised and democratised forms of communication, they are by no means guaranteed. It is equally possible that the Internet could become a means of surveillance, or become subject to state control and censorship. The current spread and use of the Internet is dominated by developed economies, where discrepancies of access relating to both socio economic status and geographical location also exist.

Although family history is an apparently benign pursuit, the uses of genealogy vary over time and place. For example, Lambert (2002) notes that the "convict stain", or stigma once felt by Australian descendants of deportees, has over the years been re-valued as a badge of honour, not of shame. There is no reason to assume that today's tastes, fashions and the use of genealogy will not change again; what has been discussed here is a recent phenomenon. Defining lines of descent, family and kinship are not only ways of claiming belonging, but in contested areas are also ways of defining ethnicity in ways that exclude others.

References

Adams, K.M. (2006), *Art as Politics: Re-crafting Identities, Tourism and Power in Tana Toraja, Indonesia*, Honolulu: University of Hawai'i Press.

Anderson, B. (1991), *Imagined Communities: Reflections on the Origin and Spread of Nationalism*, London and New York: Verso.

Bal, M. (1999), "Introduction", in Bal, M., Crewe, J. and Spitzer, L. (eds), *Acts of Memory: Cultural Recall in the Present*, Hanover, NH: University Press of New England, pp. vii-xvii.

Basu, P. (2004), "My Own Island Home: The Orkney Homecoming", *Journal of Material Culture*, 9(1): 27-42.

Beck, U. (1992), *Risk Society*, London: Sage.

Beck, U. (1994), "The Reinvention of Politics: Towards a Theory of Reflexive Modernisation", in Beck, U., Giddens, A. and Lash, S., (eds), *Reflexive Modernization: Politics, Tradition and Aesthetics in the Modern Social Order*, Oxford: Polity Press, pp. 1-55.

Beck, U. (2000), *What is Globalisation?*, Cambridge: Polity Press.

Bell, D.S. (2003), "Mythscapes: Memory, Mythology and National Identity. *Sociology*, 54(1): 63-81.
Bradley, H. (1999), "The Seduction of the Archive: Voices Lost and Found", *History of the Human Sciences*, 12(1): 107-122.
Castells, M. (2000), *The Rise of the Network Society*, 2nd edition, Oxford: Blackwell.
Castells, M. (2004), *The Power of Identity*, 2nd edition, Oxford: Blackwell.
Cattell, M.G and Climo, J.J. (2002), "Introduction: Meaning in Social Memory and History", in Climo, J.J and Cattell, M.G. (eds), *Social Memory and History: Anthropological Perspectives*, Walnut Creek, CA: Alta Mira Press, pp. 1-38.
Cavarero, A. (2000), *Relating Narratives: Storytelling and Selfhood*, London: Routledge.
Caygill, H, (1999), "Meno and the Internet: Between Memory and the Archive", *History of the Human Sciences*, 12(2): 1-11.
Chamberlain, M and Leydersorff, S. (2004), "Transnational Families: Memories and Narratives", *Global Networks*, 4(3): 227-241.
Climo, J.J. and Cattell, M.G. (eds) (2002), *Social Memory and History: Anthropological Perspectives*, Walnut Creek, CA: Alta Mira Press.
Connerton, C. (1989), *How Societies Remember*, Cambridge: Cambridge University Press.
Czap, P. (1983), "A Large Family, The Peasant's Greatest Wealth: Serf Households in Mishino, Russia 1814-1838", in Wall, R. (ed.), *Family Forms in Historic Europe*, Cambridge: Cambridge University Press, pp. 105-152.
Dandecker, C. (1990), *Surveillance, Power and Modernity*, Oxford: Polity Press.
Diamond, I. (1999), "The Census: The Biggest and Best Data Source in the UK?", in Dorling, D. and Simpson, S. (eds), *Statistics in Society: The Arithmetic of Politics*, London: Arnold, pp. 9-18.
Dicken, P. (2003), *Global Shift*, 4th edition, London: Sage.
Drake, P. (2001), Findings from the Fullerton Genealogy Study, http://psych.fullerton.edu/genealogy/.
Eriksen, T.H. (1993), *Ethnicity and Nationalism: Anthropological Perspectives*, London: Pluto Press.
Favart-Jardon, E. (2002), "Women's 'Family Speech': A Trigenerational Study of Family Memory", *Current Sociology*, 50(2): 309-319.
Foster, D. (1997), "Community and Identity in the Electronic Village", in Porter, D. (ed.), *Internet Culture*, London: Routledge, pp. 23-38.
Frances, D., Kellaher, L. and Neophytu, G. (2002), "The Cemetery: A Site for the Construction of Memory, Identity and Ethnicity", in Climo, J.J. and Cattell, M.G. (eds), *Social Memory and History: Anthropological Perspectives*, Walnut Creek, CA: Alta Mira Press, pp. 95-110.
Freeman, F. (1993), *Rewriting the Self: History, Memory and Narrative*, London: Routledge.
Giddens, A. (1971), *Capitalism and Modern Social Theory: An Analysis of the Writings of Marx, Durkheim and Max Weber*, Cambridge: Cambridge University Press.

Giddens, A. (1991), *Modernity and Self Identity: Self and Society in the Late Modern Age*, Cambridge: Polity Press.

Giddens, A. (1994), "Living in a Post-Traditional Society", in Beck, U., Giddens, A. and Lash, S. (eds), *Reflexive Modernization: Politics, Tradition and Aesthetics in the Modern Social Order*, Oxford: Polity Press, pp. 56-109.

Green, S., Harvey, P. and Knox, H. (2005), "Scales of Place and Networks", *Current Anthropology*, 46(5): 805-818.

Halbwachs, M. (1992), *On Collective Memory*, Chicago: University of Chicago Press.

Herber, M.D. (2005), *Ancestral Tales: The Complete Guide to British Genealogy*, 2nd edition, Stroud: Sutton.

Hey, D. (2004), *Journeys in Family History: Exploring your Past, Finding Your Ancestors*, London: National Archives.

Hoskins, A. (2004), "Television and the Collapse of Memory", *Time and Society*, 13(1): 109-127.

Huyssen, A. (1992), *Twilight Memories: Making Time in a Culture of Amnesia*, New York: Routledge.

Jacobson, C.K. (1986), "Social Dislocation and the Search for Genealogical Roots", *Human Relations,* 39(4): 347-358.

Jacobson, C.K., Kunz, P.R. and Conlin, M.W. (1989), "Extended Family Ties: Genealogical Researchers", in Bahr, S.J. and Peterson, E.T. (eds), *Aging and the Family*, Lexington, MA: Lexington Books, pp. 193-205.

Jedlowski, P. (2001), "Memory and Sociology: Themes and Issues", *Time and Society*, 10(1): 29-44.

Katz, J.E. and Rice, R.E. (2002), *Social Consequences of Internet Use: Access, Involvement and Interaction*, Cambridge, MA: MIT Press.

Kellerhalls, J. Ferreira, C. and Perrenoud, D. (2002), "Kinship Cultures and Identity Transmission", *Current Sociology* 50(2): 213-228.

Kitchin, R. (1998), *Cyberspace*, Chichester: Wiley.

Kraeger, P. (1997), "Population and Identity", in Kertzer, D. and Fricke, T. (eds), *Anthropological Demography: Towards a New Synthesis*, Chicago: University of Chicago Press, pp. 139-174.

Kraidy, M. (2005), *Hybridity or the Cultural Logic of Globalization*, Philadelphia: Temple University Press.

Kuper, A. (1999), *Culture: The Anthropologist's Account*, Cambridge, MA: Harvard University Press.

Lambert, R.D. (1996), "The Family Historian and Temporal Orientation Towards the Ancestral Past", *Time and Society*, 5(2):115-143.

Lambert, R.D. (2002), "Reclaiming the Ancestral Past: Narrative, Rhetoric and the "Convict Stain'", *Journal of Sociology*, 38(2): 111-127.

Le Goff, J. (1992), *History and Memory*", New York: Cambridge University Press.

Meethan, K. (2004), "'To Stand in the Shoes of My Ancestors': Tourism and Genealogy", in Coles, T. and Timothy, D.J. (eds), *Tourism, Diasporas and Space*, London: Routledge, pp. 139-150.

Meethan, K. (2006), "Introduction: Narratives of Place and Self", in Meethan, K., Anderson, A. and Miles S. (eds), *Tourism, Consumption and Representation: Narratives of Place and Self*, Wallingford: CAB International, pp. 1-23.

Miller, D. and Slater, D. (2000), *The Internet: An Ethnographic Approach*, Oxford: Berg.

Misztal, B. (2003), *Theories of Social Remembering*, Buckingham: Open University Press.

Mitra, A. (1997), "Virtual Commonality: Looking for India on the Internet", in Jones, S. (ed.), *Virtual Culture: Identity and Communication in Cybersociety*, London: Sage, pp. 55-79.

Nash, C. (2002), Genealogical Identities", *Environment and Planning D: Society and Space*, 20: 27-52.

Nora, P. (ed.) (1996), *Realms of Memory Volume 1: Conflicts and Divisions*, New York: Columbia University Press.

Osborne, T. (1999), "The Ordinariness of the Archive", *History of the Human Sciences*, 12(2): 51-64.

Pálsson, G. (2002), "The Life of Family Trees and the Book of Icelanders", *Medical Anthropology*, 21: 337-367.

Polanyi, K. (2001), *The Great Transformation: The Political and Economic Origins of Our Time*, Boston: Beacon Press.

Robins, K. and Webster, F. (1999), *Times of Technoculture: Information Communication and the Technological Order*, London: Routledge.

Skultans, V. (1998), *The Testimony of Lives: Narrative and Memory in Post-Soviet Latvia*, London: Routledge.

Sussman, G. (1999), "Urban Congregations of Capital and Communications: Redesigning Social and Spatial Boundaries", *Social Text*, 17(3): 35-51.

Tyler, K. (2005), "The Genealogical Imagination: The Inheritance of Interracial Identities", *The Sociological Review*, 53(3): 476-494.

Wang, Q. and Brockmeier, J. (2002), "Autobiographical Remembering as Cultural Practice: Understanding the Interplay between Memory, Self and Culture", *Culture and Psychology*, 8(1): 45-64.

Woods, R.I. (1996), "The Population of Britain in the Twentieth Century", in Anderson, M. (ed.), *British Population History*, Cambridge: Cambridge University Press, pp. 281-358.

PART II
GENEALOGY AS A CULTURAL PRACTICE

Chapter 7

Genealogical Mobility: Tourism and the Search for a Personal Past

Dallen J. Timothy

Introduction

Tourism is often said to be the largest industry in the world. Whether or not this is true is a matter of debate, but tourism is in fact one of the most significant socio-cultural and economic forces today, and no place on earth remains untouched either directly or indirectly by tourism. Every year hundreds of millions of people travel to experience cultures, climates, and natural landscapes different from those at home. Geographers and other social scientists have identified a significant range of tourism types based upon destinations and attractions visited, kinds of activities undertaken, and motives for travel. Among the most commonly cited categories of tourism are nature-based, sport, religious, beachfront, business, cultural, and adventure. Cultural heritage tourism is among the most pervasive forms of travel and includes both tangible and intangible elements of the human past as its resource base. Heritage specialists define heritage as "the modern-day use of the past" for a variety of reasons, including conservation, education, entertainment, and tourism (Ashworth, 1995; Graham et al., 2000; Hewison, 1987; Timothy and Boyd, 2003). Heritage resources for tourism typically include museums, literary places, historic buildings, villages, cultural landscapes, monuments, event re-enactments, battlefields, castles, churches, traditional cuisines, folkways, folklore, sacred places, and archaeological sites.

Heritage tourism occurs on a number of different scales, depending on the sites being visited and the tourist's degree of personal connection to the place. Historic places such as Stonehenge (England), the Pyramids of Giza (Egypt), Machu Picchu (Peru), and the castles and cathedrals of Europe are well-known international heritage resources that draw millions of people each year from around the world. However, many lesser-known places draw fewer spectators but are no less important in the whole of human heritage. These typically occur at the scale of national and community heritage (Timothy, 1997). At perhaps the lowest scale of heritage are people, places, artifacts, and events that are important to the lives and identities of individuals and families. The past at this level is a highly significant tourism resource, even though it is far more difficult to define, delineate and measure. Nonetheless, the quest for personal heritage is one of the most salient resources and motives for travel today as people in the developed world become increasingly aware of their personal pasts through genealogy and family history research. Often the quest for roots motivates people to travel to their ancestral lands in both domestic and international contexts. Genealogy and its broader cousin, family history, are a rapidly growing pastime among many segments of modern society; accompanying this growth is a recent surge in travel for personal heritage reasons. Much of this growth can be attributed to increasing interest in genealogy, but it is also influenced by sheer interest among diasporic peoples to experience the lands of their ancestors.

This chapter builds upon existing knowledge about heritage tourism by examining one of the most pervasive yet least understood elements of heritage tourism, namely personal roots tourism and its specific component, family history travel. The chapter first examines global diasporas and the heritage identity crises that have accompanied these migrations and cause people to yearn for the past. It then examines the phenomenon of personal heritage tourism, or roots/family history tourism, as a specific form of heritage tourism and provides an overview of current trends, as well as many of the experiential aspects of demand and the functions of supply in providing the personal heritage travel product.

Diasporas and Rootlessness

Diaspora refers to nations of people living outside their traditional homelands (Cohen, 1997; Mitchell, 1997). Coles and Timothy (2004: 3) refined this as "groups of people scattered across the world but drawn together as a community by their actual (and in some cases perceived or imagined) common bonds of ethnicity, culture, religion, national identity and, sometimes, race". Throughout history, ethnic groups have migrated abroad by various forces and for a variety of reasons. Cohen (1997) suggested a fivefold classification of these movements: Victim Diasporas, Labor Diasporas, Imperial Diasporas, Trade Diasporas, and Cultural Diasporas, each with its own set of mitigating circumstances and degrees of personal freedom in choosing whether or not to leave.

Cohen (1997: 26) identified nine common features of diasporas that have a significant bearing on this chapter; he also recognized that not all diasporas possess all of these characteristics:

- Dispersal from an original homeland, sometimes traumatically, to two or more foreign regions.
- Alternatively, departure from a homeland to seek work, trade, or to advance colonial aspirations.
- A collective memory and myth about the homeland, including its location, geography, history and achievements.
- An idealized regard for the homeland and a united commitment to its maintenance, restoration, safety, and prosperity.
- The development of a return movement that gains collective approbation.
- A strong ethnic group consciousness sustained over a long period of time and based on a sense of distinctiveness and common history.
- A troubled relationship with host societies, suggesting a lack of acceptance at the least.

- A sense of empathy and solidarity with co-ethnic members in other countries of settlement.
- The possibility of a distinctive creative, enriching life in host countries with a tolerance for pluralism.

Many diasporic groups reside and function in a host country but retain strong emotional and allegiant connections to their countries of origin (Kelly, 2000; Sheffer, 1986). For people of relatively recent migration, and perhaps even for many multi-generational descendants, this loyalty and bond to the idealized motherland gives rise to "halfway populations" (Hollinshead, 1998) that often express, perhaps even unbeknownst to them, feelings of unease, ambiguity, restlessness, and confusion about their interstitial position in their adopted land (Coles and Timothy, 2004; Louie, 2003; Richards, 2005). Thus, it might be said that diasporic identities are creolized, hybridized or hyphenated (Friedmann, 1999; Hannam, 2004; Soja, 1996), referring to semantic terms such as African-American, Indian-British, Lebanese-Canadian and Irish-Australian—titles that identify people by their association with the old and new homelands. Many people in today's western societies, however, are "mongrels" of sorts, comprised of many mixed ethnicities and diasporic lineages, so that such clear-cut hyphenations and identities are less relevant. Nonetheless, the indicators of "in-betweenness" often placed upon migrants and their progeny by others, is a manifestation of forces that play a salient role in the creation of diasporic identities and their people's connections to the spaces of their past and present. Thus, a crisis of discontinuity may be created, causing some people to wonder who they are, where they came from, and stirring within themselves a desire to be complete. Basu (2005a: 124) articulated it well as "a crisis in belonging that has been characterized as a post-colonial unsettling of settler societies".

This crisis in belonging does not affect all immigrants and their progeny. Many immigrants are happy to assimilate into their new homeland, and there is evidence to suggest that this identity challenge likely diminishes with subsequent generations (Card et al., 1999; Nauck, 2001). For many people in mainstream America, for instance, there is very little left of the diasporic identity, particularly among people whose ancestors emigrated so long ago that any ties with the "homeland" are either lost or newly-minted fabrications either for tourism purposes (e.g. homeland souvenirs sold at cultural festivals in the United States) or for political motives (e.g. Quebéc's ties to France). While many of these people may be passionate about finding their roots, they are not necessarily "restless" in the new world, continually longing for their historic homeland.

There is also an element of race involved in this, for it is easier for non-visible minorities to assimilate into host societies (Hou and Balakrishnan, 1996). "Whiteness" can for example, in Europe, North America, and Australia confer a sense of being "in place", because of racist social undercurrents.

Rootlessness is apparently only visible in diasporic and colonial societies, as in indigenous and traditional societies "there are no disputes about [their] origins" (Hall, 2007: 1139) and "their history seems as much a part of their lives as eating, sleeping, shopping, and going to work. They know who they are because they know where they come from. They acknowledge and thank those on whose shoulders they stand, who passed on their genes, culture and wisdom. It [is] as unthinkable to neglect the ancestors as to neglect to dress or comb one's hair" (Fein, 2006: 39). For example, among the Maori of New Zealand, Fein (2006: 37) was

> stunned to find that every Maori I met knew the name of the canoe his ancestors had arrived in, over a thousand years ago. A Maori historian...recited his lineage back across the great oceans, for thousands of years...Knowing who he was and where he came from gave him a foundation so solid that it seemed nothing could knock him over...when someone asks us Americans about ourselves, we tell them our names, of course, and, often what we do for a living...Our universe contracts when we identify ourselves only by our current lives...we have nothing to lean on, no permanent sense of place, no connection to what came before us. And when our children and grandchildren...are gone, it will be as though we never existed.

Such a complexity of emotions and identity in a hyphenated life, even among some people whose ancestors migrated many generations earlier, often manifests in everyday life as a sense of nostalgia (Akhtar, 1999; Nguyen and King, 2004). Nostalgia implies a yearning for some past socio-spatial condition(s) and evokes bittersweet wistfulness within individuals and entire societies shared by people of similar backgrounds, such as cultures, nations and generations (Belk, 1990; Davis, 1979). Such sentiments often suggest that life in the distant past and in the fantasized and idealized "motherland", "homeland", or "old country", was more wholesome, meaningful, hallowed, and sincere, compared to the superficial, mechanized, materialistic, mass-produced, depersonalized, and harried lifestyle in contemporary western societies (Lowenthal, 1996; Nash, 2002), particularly North America, in the same way that Bunce (1994) suggests some people sense the "countryside idyll", where warm feelings, a sense of morality, and positive images are typically associated with rurality. Likewise, Lowenthal (1979) argued that, in conjunction with rapid development, the modern-day destruction of historic buildings, places, relics, and traditions has deepened people's nostalgia for the past. He suggests that a search for roots and historical identity and an increased appreciation for one's familial patrimony is evidence of this trend. Travel to ancestral homelands is one of the fastest growing contemporary manifestations of this identity-seeking trend.

Personal Heritage/Roots Tourism

In recent decades there has been a surge of interest among Americans, Canadians, Australians, Chinese and many others of diasporic descent in discovering their personal heritages through family history and genealogy research. Some people believe that they must understand the past in order to understand their present (Basu, 2004a; Kurzweil, 1995; Lowenthal, 1975), as is evident with current interests in genetics and genealogy. An informant in Basu's (2004b: 166) study of family history tourists in Scotland explained, "I come from those people, I share that psyche, therefore journeying into it is a journey into myself". This interest in personal pasts has led increasing numbers of people to travel to ancestral lands in an effort to resolve their personal identities, connect emotionally with their deceased predecessors, and feel at home on their "native" soil. Personal heritage is the least understood scale in heritage studies from an academic perspective, probably because it is more elusive than other forms and the experience more nostalgic, personal, subjective, and sometimes spiritual; thus, it is harder to define and evaluate.

In the words of Kevin Lynch (1972: 40), "most Americans go...to Europe to feel at home in time", because, with some notable exceptions (e.g. re-built structures whose originals were destroyed in World War Two and modern urban high-rises), much of the human environment of Europe is ancient and durable—something lacking in the rapidly evolving cultural milieu of North America, where landscapes of the ordinary are being replaced by landscapes of the grandiose and post-postmodern. Thus, traveling to distant homelands may help people explain and evaluate themselves from many perspectives, including health, physical appearance, personality, and individual habits. For some people, diasporic "return travel" is important for them in understanding the context of their upbringing, given family behaviors, patterns, and relationships related to their parents, grandparents, or even earlier predecessors. In the words of Lowenthal (1975: 6), "the past gains further weight because we conceive of places not only as we ourselves see them but also as we have heard and read about them". This sentiment sunk deep in the experience of one genealogy tourist who visited the childhood home of his father. "For all the times my father told me about Shanghai in my 30-plus years, my feel for it was much like...an image that didn't quite seem real. I'd heard about the city so often, that it became almost mythical to me, until...I stepped back in time and saw it for myself" (Compart, 1999: 1).

Roots tourism focuses on places that have direct connections to an individual's own lineage or familial legacy. Visiting personal heritage locations and undertaking family history-related activities are generally very different from the experiences of general heritage consumers who visit historic sites on an individual day trip or organized tour, owing to the different levels of personal and emotive connections to place and artifact (Stephenson, 2002; Timothy, 1997; Timothy and Boyd, 2006).

Family history research and genealogy are important aspects of the personal heritage tourism phenomenon. In Salt Lake City, Utah, the Church of Jesus Christ of Latter-day Saints (Mormons) operates the world's largest genealogical library. Thousands of people travel to the center each year to investigate their family histories and to try to unravel their personal and familial identities (Olsen, 2006). According to an early 1990s survey (Hudman and Jackson, 1992), a quarter of all visitors to Temple Square in Salt Lake City came for genealogical purposes. Popular guidebooks have been published to encourage visits to the Genealogy Library and to simplify the research process (e.g. Parker, 1989).

Meeting distant and close relatives and attending family reunions are other important types of genealogical experiences that can in some cases entail traveling great distances. This ranges from small gatherings of immediate family members in home towns to enormous assemblies of entire clans in faraway lands (Asiedu, 2005; Duval, 2002; Feng and Page, 2000; Legrand, 2005; Neville, 2003; Sutton, 2004). Often these are important venues for exchanging ancestral information, photographs, and relevant stories and serve to reaffirm kindred connections wherein one is part of a large entirety of interconnected people (Sutton, 2004).

Other activities involved in family history tourism include research in community and church archives, collecting data from gravestones, photographing old family farms, collecting or copying documents associated with the lives of ancestors (e.g. marriage certificates), visiting museums, purchasing period antiques, touring family or clan-lands, and visiting places of worship associated with special events in the lives of forebears (Horst, 2004; Timothy, 1997; Timothy and Boyd, 2006). Churches play an especially important role in family history travel itineraries for at least three reasons. First is because they may contain original parish registers that are accessible in no other way. Second, in the case that ancestral houses or farmsteads are unknown or no longer exist, the church is generally the most durable part of the community and can be visited and photographed as a representation of the original community. It likewise provides a sense of shared space with visitors' ancestors. Finally, there may also be a religious component if descendants still practice the same faith their forebears practiced.

Genealogy-based travel is the epitome of personal heritage tourism. It provides opportunities to explore a side of personal heritage that cannot be understood or felt at home. During travel to ancestral places, opportunities are often available for easier archival data collection, interviews can be done more easily, and photographs can be taken. Less pragmatic, but equally important, if not more so, travel allows family historians to experience first hand the homes and environments that their ancestors knew "in the bosom of the Mother Country" (Louder, 1989: 136). Often this encounter has strong spiritual and emotional connotations. "One of the most exciting experiences a genealogist can have is to visit and research in the very place and area where their (sic) ancestors lived" (Bradish and Bradish, 2000: 44).

In the phenomenon of personal heritage tourism many players are involved in creating the roots experience. The following sections highlight three of these: the travelers themselves, tour and service providers, and government agencies that oversee tourism in destination countries.

Roots/Family History Tourists

It is understandable that amateur and professional genealogists would have a desire to travel. Many enthusiasts report that doing genealogy and family history draws people closer to their ancestors (Dunne, 1989; Kurzweil, 1995; Lambert, 1996). After beginning family history endeavors or exhausting all known resources in one's home community or region, for many people the natural next step would be to visit family history libraries away from home, old homesteads and towns within their own countries, and ultimately trips abroad to visit the homelands of their predecessors (Basu, 2004b). Geographers and other social scientists still know little about people's deeper purposes for traveling to their ancestral lands, their experiences while there, and their reactions and interactions with what many consider "sacred personal space". However, some scholars have begun to research this extraordinary set of forces at play in the family history tourism phenomenon with interesting results (e.g. Basu, 2004a, 2005b; Du, 2006; Kelly, 2000; Keng, 1997; Meethan, 2004; Nash 2002).

One important expression of restlessness and rootlessness in the adopted land is a desire to connect with ancestral places. According to Tuan (1974), topophilia, or the love of the land, is an intrinsic human sentiment that is most observable in traditional societies where group identities are firmly rooted in the native soil. Many examples exist today of this among refugee groups that have been displaced from their traditional homelands. On the island of Cyprus, for instance, once the ban on cross-border travel between north and south was lifted in 2003, thousands of Greek Cypriots traveled north and Turkish Cypriots traveled south across the Green Line to visit the lands they were expelled from in 1974; many since that time have undertaken the journal many times because of their strong fidelity to the land (Dikomitis, 2004; Webster and Timothy, 2006).

After arriving at the familial destination, personal heritage tourists commonly comment on how they feel a natural bond to the land (Tuan, 1974; Nash 2002), expressed by Basu (2001: 343) as being caused by the "blood of your ancestors stirring within you". Matthiessen (1989) calls it "primordial longing" for the homeland, which manifests in many people's minds as an idealization of the sacred space of the homeland. By traveling, setting foot on the soil, seeing, touching, and smelling ancestral terra, ties are strengthened to a particular territory, clan-lands full of stories, symbols, landscapes, and myths. Participating in ceremonies, festivals, clan marches, and reunions, which some destinations offer, may help strengthen ties to the homeland (Basu, 2005a; Hannam, 2004). However, many of these are staged tourist spectacles performed in languages visitors do not understand, thus contributing little to some people's connection to place and possibly even distancing them from their ancestral lands as visitors feel disappointed in being treated like ordinary "tourists".

This indicates that individual experiences vary from person to person and locale to locale. Nonetheless, visiting lends narrative and can potentially make real the natural and cultural landscapes described in historical family records but most often only imagined by the contemporary family historian. Thus, the materiality of this homeland "empowers the experience, making it…"real" [to the personal heritage tourist] and therefore essential in terms of identity" (Basu, 2004a: 40). In Scotland, geographical tools, such as clan maps and family museums (e.g. Clan Macpherson Museum in Scotland), together with tours that highlight the hillocks, streams, valleys, and fields known to one's ancestors, are often utilized by roots destinations to enhance this sense of belonging to the land, thereby creating a sense of solidarity as people re-root their identities with others who inhabit or have inhabited these "sacred places" (Basu, 2004a).

Similarly, many roots tourists describe increased connections to their deceased forebears. A study by Timothy and Teye (2004: 119) found that many African American roots tourists feel very connected to their ancestors. "Some people felt as though their progenitors were speaking or crying to them…Others commented on the pain they felt for their ancestors… 'I almost broke down and cried for their pain'". Kemp's (2000: 12) own roots journey caused her to come to terms with her ancestors' plight:

> I don't find it disturbing. At least not any more. Because I understand that I have been singled out by the spirits of my ancestors to tell their stories. It is their presence I feel

beside me and it is why they whisper incessantly from the moment I arrived in [Ghana], "where have you been daughter? There is work to do".

Similarly, Jorgensen (1988: 23) explained how

> It is impossible to describe the emotions of love and kinship I have felt for my ancestors as I have visited the areas where they lived. A treasured moment in my life came when, on a trip to Sweden, I came upon a fallen-down stone building. Somehow, I knew I was standing on sacred family ground. Later research confirmed my impression—those stone ruins had been a chapel in which some of my ancestors were married.

During trips to the old country, people are often drawn to specific individuals they have researched before leaving home. Frequently, they arrive home from the journey with a better understanding of their ancestors; the heritage pilgrims come to "know" their departed at a higher level than before (Basu, 2004b). For many, "it's a very emotional thing. Our family has always been very important to us, to see where our earliest recorded ancestors actually worked and made a living, and to see the things that they saw, looking over the bay, it's…very emotional" (Basu, 2004a: 37).

Family history is a spiritual, or even a religious experience for many participants (Jorgensen, 1988; Lambert, 1996, 2006; Olsen, 2006; Thorstenson, 1981). Among the most active genealogists are members of the Church of Jesus Christ of Latter-day Saints, who are obligated for religious reasons to carry out genealogy research. LDS theology involves the interconnectedness of generations (Bushman, 1980). For most Mormons, however, genealogy is more than a doctrinal duty; it is a pleasurable pursuit, a faith-promoting experience, and a deeply enriching, spiritual endeavor: "Spiritual experiences happen almost daily to those involved with this work" (Rodriguez, 1987: 12).

Spiritual feelings and theological obligations lead some genealogy travelers to shun the title of "tourist" and adopt the persona of "pilgrim", for they see themselves not as pleasure-seeking hedonists but as pilgrims on a sacred quest for spiritual enlightenment, peace, and identify confirmation (Basu, 2004b). Many African Americans describe their roots tours in Africa as very sacred experiences. Faith in God is renewed for those who rely on their beliefs to deal with the pain of visiting the sites of horror associated with slavery. Some report feeling that the soils of Africa are sacred ground where God walks with the spirits of their dead (Timothy and Teye, 2004).

Roots travel can be a cathartic experience among populations whose ancestors were systematically brutalized by people in power or otherwise imprisoned and tortured by colonial authorities, such as Jews and descendants of African slaves. For these people, ancestral heritage tours to places of traumatic memory provide opportunities to heal and gain closure to historical transgressions that affect them still today (Schramm, 2004; Timothy and Teye, 2004). Others travel to their ancestral lands to get away from mistreatment in their own country. "In black American communities…we journey to Africa, hoping to briefly escape American racism and experience racial dignity at its source" (Richards, 2005: 620). Despite these pure sentiments, much to the disappointment of many Americans, Caribbeans, and Europeans of African descent, they may be treated by their African counterparts not as insiders but outsiders—simply foreign tourists who only belong in Africa if they have money to spend. This form of prejudice adds insult to injury for many travelers, who already suffer ancestral identity crises. This rejection by their propitious "brothers and sisters" in Africa digs into extant emotional wounds and creates a deeper sense of liminality in their lives and feelings that they do not belong in their current homeland and are not welcomed as kin in their motherland (Bruner, 1996; Clarke, 2004; Hasty, 2002; Timothy and Teye, 2004).

These experiences associated with connections to place, kin, and spiritual sensitivities sometimes, as several authors have noted, result in a desire among genealogy tourists to consume their ancestral space either symbolically (visiting, photographing, gazing) or literally. For example, in writing about a couple from Arizona (USA) visiting their ancestral lands in Scotland, Basu (2004b: 150) recorded that after learning about the details of their ancestral place, they were shown an old well that had once served their ancestral village. "With tears in their eyes, they knelt first to cup the still-clear water in their hands to drink, and then to fill their emptied flasks so that they could carry the spring water back to Arizona with them". Schramm (2004: 146) noted similar behaviors among African American visitors to their ancestral lands in Ghana, with one in particular drinking water from a river.

> At first he did not want to do that—having in mind all the warnings about unwholesome water and terrible diseases associated with it. But then he gave it a second thought... maybe my great-great grandmother or grandfather took their last bath on African soil in that river...so I reached down and got the water and drank it, so that always a part of me would be the last bath of my ancestors before they were taken away from Africa.

Likewise, genealogy tourists commonly collect relics from ancestral places as a way of preserving their experience and laying claim to the lands of their predecessors. Stones, leaves, soil, pieces of wood, pottery fragments, river or well water, and antiques are considered "sacred substance" and can be displayed prominently on their mantels-cum-sacred-shrines at home (Basu, 2004b). Likewise, enjoying local foods in restaurants or bringing home local foods and cookbooks are part of the literal consumption of the place as well.

For similar reasons of connectedness, some roots pilgrims/tourists are also known to leave something of themselves in familial homelands. According to one tourist, who placed her ring among the loose stones of a wall near her ancestral home, "by putting my cheap ring inside the walls, I felt I was giving a humble offering...almost begging to be part of it forever...I wanted to leave a part of 'me' there...when I did it, I felt extremely good, extremely relieved" (Basu, 2004b: 167).

Ancestral Tour and Service Providers

Between the roots tourists themselves and the governments of the countries they visit are situated many layers of service providers and intermediaries, although many people organize their own genealogy trips. Primary among service providers are tour operators, professional family historians, accommodations and food service suppliers, and transportation providers. African diaspora tours tend to be comprised of large groups, owing to the difficulty in traveling in much of Africa on an individual basis and the traditional lack of genealogy libraries, archives, marked graveyards, and other elements of most ancestral tours. The primary purpose of African roots tours is to provide opportunities to tour various regions, most notably in Ghana, The Gambia, and Nigeria, to see how the countryside looks, to try local foods, to visit slave trade-related sites (e.g. slave forts and monuments), and meet African people who may or may not be distant blood relations (Bruner, 1996; Essah, 2001; Jobe, 2004). European roots tours are usually conducted in small groups or on an individual or family basis, because they typically involve more in-depth family history work, such as archive searches, cemetery visits, and home site visits—activities based on resources that are better preserved than in Africa and many other less-developed regions.

Tour companies or destination management organizations (DMOs) are perhaps the most notable of these service providers, as they assemble packages that include the sub-contracted services of others, such as lodging and genealogy specialists. Diasporic genealogy and cultural societies, such as the Toronto Ukrainian Genealogical Group and the Polish Genealogical Society of Connecticut and the Northeast, Inc., also function as tour organizers. There are literally hundreds, if not thousands, of genealogy tour brokers throughout the world attempting to cash in on the growing desire among diasporic peoples to discover their ancestral roots. These companies cleverly utilize phrases that touch people's sense of longing for links to familial legacies, spiritual connections to deceased progenitors, nostalgic sensitivities for the simpler life of yesteryear in the Old World, and understanding of self through tradition and birthright bonds. For example,

> Here, you can take yourself off from the normal tourist trails and discover your own unique connection to Ukraine. Visit the towns and villages where your ancestors lived, the churches which they worshipped in and the fields and hills in which they walked. You will definitely feel that you are stepping back to the time of your ancestors...Bring your past into the present and gain a fascinating insight into the lives of your ancestors. This is not only a trip of a lifetime, but a trip of lifetimes! (Toronto Ukrainian Genealogical Group, 2007)

> Let us organise an incredible week for you...as we take you where your ancestors lived and worked and following in their footsteps. Weaver, coal miner, farmer or laborer—let us show you the villages and towns where your kin-folk walked and dreamed of dreams. (Ancestral Journeys of Scotland, 2007)

> Imagine the thrill, the excitement when you discover where they lived and worked. What made them leave the land of their forebears? Can you sense the excitement when you read actual newspapers that your ancestors may have read? Many of these newspapers carry advertisements offering a new life in foreign climes. (Scottish Genealogy Tours, 2003)

A thorough examination of online and printed family history tour products offered by various agents in Europe and Africa (see Timothy, 2001) reveals several common products or services that genealogy tour operators offer to family history-based visitors. The first is an orientation to the ancestral region. In this case, tourists are taught about the landscape, climate, economic conditions, and cultural trends of the broader region as they were in earlier times and in the present. How these regional characteristics affected the lives, activities, and thinking of the area's inhabitants is a central component of this orientation, particularly as it pertains to understanding people by understanding the region where they lived. Visitors tour villages and provincial capitals, participate in area celebrations, and walk through the rural and urban landscapes where their ancestors might have walked.

Second, providers orient visitors to the ancestral community by walking them through their ancestral village, "breathing the same air their ancestors breathed". Seeing the roads, buildings, churches, cemeteries, historical sites, monuments, and farms that were an everyday part of their ancestors' lives is an important part of the product as well. Some people are even given the chance to attend Sabbath services in a house of worship where their forebears were christened or married.

Another service is helping family historians locate and visit their ancestral homes. Many European communities take great pride in, and devote considerable efforts to, maintaining and preserving old buildings and historic homes. It is not uncommon today for people in rural England or Germany, for example, to live in homes that are several centuries old. Many pre-packaged tours require information ahead of time, so that on-the-ground staff can locate personal heritage places in advance. If ancestral birth houses still exist, tour companies may contact current owners to request permission to visit and tour the home and farm. Related to this is the service of providing copies of ancient land registers and cadastral maps to assist tourists in identifying their familial properties correctly (Ancestral Roots Travel, 2007).

Educating diaspora descendents about their ancestors' everyday life is another typical part of the roots experience. It sometimes involves showing and demonstrating appropriate period work accoutrements, farming and other work techniques, and styles of food and dress. Many museums exist today that depict the ordinary lives of common folk with artifact displays and live actors and demonstrations. Information about the lives and work of miners, farmers, blacksmiths, millers, weavers, and other trades, assists customers in appreciating their personal pasts and understanding better the lives of their progenitors.

The fifth service is introducing visitors to distant relatives and providing interpretation. Meeting distant relations is often arranged so that overseas visitors can talk and exchange photographs, experiences, and addresses (Genealogy Germany, 2007; Thorstenson, 1981).

Hands-on family history and genealogy research is another important service that tour operators provide, typically with this work contracted out to family history consultants. If desired, time is scheduled into itineraries to search archives, church records, genealogy libraries, cemeteries, civil documents and tax rolls. Often the tour companies have specialized staff members who research relevant sources and archives in advance of tourists' arrivals (European Focus, 2007; Hooked on Genealogy Tours, 2007; Select Travel Service, 2007).

Many operators also assist clients in trying "authentic" local foods. Customers can eat meals with local families, prepared in traditional ways with local ingredients, and even participate in preparing the meals. Again, this assists family historians in connecting in another way with their ancestors through heritage cuisines—typically peasant foods that would have likely been the staples of earlier days (Genealogy Germany, 2007).

Finally, family history tour companies are able to arrange accommodations, transportation, shopping trips and other activities, such as golf or sailing. While many genealogy tourists probably feel hurried for time, others may desire to take an occasional afternoon off from the rigors of "doing" roots. Some people may also "drag along" travel companions who are not as enthusiastic about genealogy and might prefer a day of golf to a day in the archives. Such added amenities and services contribute additional income to the basic elements of ancestral tour organizers.

Government Agencies

Nearly every country in the world has a national-level public agency that oversees the research, development and promotion of tourism. The United States is the primary, and most noteworthy, exception to this rule, although individual U.S. states have their own official tourism promotion bureaus.

In their role as promoters of place and creators of image, national tourism organizations (NTOs) are responsible for building awareness of the tourism opportunities in their countries for visitors and foreign investors. In addition to their responsibilities related to human resource training, service quality assurance, and data collection, they must also drive the design of brochures, posters, and web sites and actively extol the cultural and ecological virtues of their homelands at international trade fairs among global distributors.

Tourism is big business throughout the world with significant economic ramifications. As a result, destination countries and regions are engaged in a constant search for new markets and new product-based competitive advantages. In the realm of personal heritage tourism, NTOs of many countries that have experienced large-scale emigration during the past few centuries are actively involved in promoting "return visits", "roots routes", and other forms of family history-based tourism (Hungarian National Tourism Office, 2000; Magenheim, 2002; Morgan and Pritchard, 2004; Wales Tourist Board, n.d.), and many of their official web sites devote significant attention and space to the salient relationship between tourism and family history. Examples of these efforts can be found in all parts of the world, including Asia, Africa and Europe—the origin regions of most of the world's diasporic communities.

The Irish Tourist Board (Fáilte Ireland) has actively promoted genealogy-based tourism for many years. It regularly assists in organizing sept gatherings that attract tens of thousands of people to Ireland each year. In the 1980s and 90s, with the economic power of tourism as a driving force, over 250 such reunions were typically planned each year by Fáilte Ireland (then known as Board Fáilte) as part of a recent surge of heritage-related projects designed to attract overseas interest in documenting and participating otherwise in Irish family history. "Seemingly overnight, what was once considered a lure attractive only to genealogy hobbyists has emerged as an ambitious government-assisted project of science, history, and tourism" (Keating, 1990: 46).

Similarly, Visit Wales (*Croeso Cymru*, formerly the Wales Tourist Board) organizes mass reunions for Jones, Griffith, and other prominent families of the Welsh diaspora, and Welsh genealogy features prominently on its web site (http://www.travelwales.org/) (Bly, 1998). Likewise, as part of the "It's time to Come Home" (*Mae'n Bryd I ddod Adref*) campaign aimed at the Welsh diaspora, the Wales Tourist Board published guides and handbooks to assist visiting Welsh descendants in doing family history in Wales (Wales Tourist Board, n.d.—see also http://www.homecomingwales.com/findingfamily.php), TV ads and various promotional videos to appeal to the hyphenated Welsh community (e.g. Welsh-American) with feelings of identity, nostalgia and connectivity to the motherland (Morgan and Pritchard, 2004). Many examples of similar efforts exist in places such as Scotland, Poland, Germany, Hungary, Switzerland, Italy, and many other sources of European diasporic kin. The Hungarian National Tourist Office's (2000) successful "Routes to your Roots" campaign brought large waves of Hungarian offspring from around the world, particularly the United States and Canada, to participate in genealogy tours with such beguiling and emotionally imbued themes as "Hungarian Heritage Discovery", "Retrace your Hungarian Roots", "Visit the Hungary of Today and Meet its Past", "Out of the Melting Pot", and "In the Footsteps of Jewish Emigrants"—tours where Hungarian progeny could "…discover your ancestral home, the town and the streets that…your family came from" (Hungarian National Tourist Office, 2000: 4-5).

While economics plays a crucial role in these promotional efforts, there is also a strong political element associated with such roots tourism campaigns. Some governments have capitalized on family history tourism to encourage political support, sympathy for various political causes, and to build a sense of nationalism and pride not only in residents of the country but also among members of their diasporas (Sheffer, 1986). A good example is Croatia, where the government encouraged visits by diasporic Croatians as a way of garnishing support for independence from Yugoslavia, to fund political endeavors, humanitarian aide, and elections. Travelers from the Croatian diaspora were encouraged by the Croatian state to play several importance political roles in the periods of conflict and post-war recovery of the 1990s: soldiers, interpreters, political advisors, and business investors (Carter, 2004: 193-194).

Although diasporic youth travel is not genealogy-based tourism in the strictest sense, it is marketed as a "homecoming" and should also be mentioned in this context. Several countries, including Israel, China and Croatia, have initiated programs designed to bring youth of their diasporic fold "home" to learn to appreciate the country's history, cultures, and politics, and to develop loyalty to the broader national cause (Carter, 2004). While not all of these programs are government sponsored, most of them are supported and funding largely by national governments and initiated by public-sector agencies in conjunction with cultural groups and philanthropic associations. The Chinese Youth League of Australia and the China Youth Travel Service are two examples of agencies that take ethnic Chinese youth from around the world to visit significant places in China that represent what it means to be Chinese (Louie, 2003). This is a significant part of the larger movement of Chinese compatriots from around the world visiting their ancestral homeland, to which all Chinese, regardless of place of residence and social integration, are expected to be a part (Keng, 1997; Lew and Wong, 2004). Similar programs have been established by the government of Israel in cooperation with Jewish associations around the world, including "Israel Experience" and "Birthright Israel" (Cohen, 2006). Birthright Israel is one of the best known of these and focuses on bringing Jewish youth (under age 26) to Israel on peer education trips to experience language, culture, sacred sites, and religious ceremonies in an effort to garner support for the Israeli cause among the rising generation of global Jews. Unlike most programs, the cost of Birthright trips is covered by the state of Israel and other supporting organizations. To date over 120,000 young Jewish people have participated in this program since its inception in 2000 (Birthright Israel, 2007).

Conclusion

This chapter has examined the complex dynamics of family history or roots tourism, with special emphasis on family historians, and more broadly diasporic peoples traveling to their ancestral nations to connect to the land itself and the spirits of their deceased antecedents. While most of the experience of personal heritage tourism involves touring natural and cultural landscapes, including ancestral villages and towns, the most devout family history travelers (from a European perspective at least) spend much of their time delving into their personal pasts, getting to know more about their predecessors via archival research, cemetery visits, meeting distant relatives, attending extended family reunions, and visiting homesteads and religious spaces associated with the lives of their departed.

It is evident that different ethnicities and cultural groups, as well as different people, have dissimilar experiences in their quests for familial understanding. This is even pragmatically obvious in the compositions and consumptions of ancestral tour products themselves. Destinations that are less equipped with archives, demarcated homesteads preserved through cadastral mapping and other landholdings records, easily identifiable modern-day relatives, and marked and documented cemeteries, are less able to provide the experience that places in Europe and parts of Asia are able to offer. Thus, trips to western Africa, for example, offer fewer of these specific experiences, but concentrate more on meeting people, touring regions where ancestors could possibly have originated, and visiting museums and monuments, such as slave trading forts and slavery memorials. Regardless of the degree of depth in ability to conduct genealogy or family history, all roots tourists appear to have several things in common. These include a personal attachment to the lands of their ancestors and bonds with their progenitors; many also report being spiritually moved.

Although this essay has taken a Euro-normative and post-settler society view, as noted by the references in this chapter to Chinese and African diasporas, the appeal of personal heritage extends far beyond modern Europe. Additional attention must also be given to Asian, African, Middle Eastern, Latin American, and Pacific Islander diasporas and their descendants' quests for family history knowledge and affirmation through travel. This need is especially salient in light of the diverse historical forces at play that caused large ethnic populations to leave their native lands in the first place (e.g. slavery, poverty, war, work, family unification, colonialism). Although relatively little is still known in the tourism context about the influences of these variables on subsequent generations, there must be little doubt that forced migrations were hugely divisive among first-, second-, and third-generation visible minority migrants and indigenous people, particularly among peoples who were brutalized by colonialists and other intruders. These people were (and are) more visible targets of racism and other forms of bigotry, and some their offspring have expressed a considerable need to come to terms with the past through the cathartic intensity of family history travel. In this regard, the important cultural, political, and social roles of the ancestral land, the new land, and the process of migration cannot be overstated in the mediation of diasporic identities (Ali and Holden, 2006; Anthias, 1998; Bhatia and Ram, 2001; Coles and Timothy, 2004; Gordon and Anderson 1999; Gupta and Ferguson, 1992; Mitchell, 1997; Safran, 1991; Smith, 1999).

Similarly, little is known about roots tourism in the domestic context. For many descendants of immigrants this is the only viable travel alternative, in terms of financial and lineage knowledge constraints. Such domestic movements are obviously important and include indigenous people returning home and diasporic peoples visiting the communities and homes of their ancestors. Many African Americans travel each year from around the United States to visit the places where their ancestors toiled, suffered, and lived in the South (Bartlett, 2001; Hayes, 1997; Woodtor, 1993). Slavery-based heritage is an increasingly important part of the United States tourism industry, not only for black Americans, but for all Americans, since more truthful histories are only just beginning to be revealed in that region (Timothy and Boyd, 2003). Domestic ancestral travel may be as rewarding, cathartic, and spiritual as international travel, particularly for genealogists with long pedigrees within their own countries.

A final perspective is the notion of two-way travel. Diasporic children are not the only ones traveling for diasporic reasons. Many examples exist where present-day citizens of diasporic source lands travel the world to experience the transplanted cultures of their ethnic brothers and sisters. Icelanders visit Icelandic settlements in Utah (Thorstenson, 1981); Germans travel to Brazil and Venezuela, and Welsh to Argentina to see those countries' German and Welsh settlements (Ferguson, 1995; Gade, 1994); Finns journey to Florida and the Upper Great Lakes region of the U.S. and Canada to visit their compatriot communities (Timothy, 2002); and many French Canadians visit the ever popular "Floribec" Quebécois region of Florida (Tremblay, 2006). All of these perspectives, domestic roots tourism, racism, experiential differences between groups, etc, could be fruitfully brought to bear on future personal heritage tourism research.

The rapidity of technological growth means an increased potential for people who cannot afford to travel to ancestral lands, to become more integrated in virtual space into their online ethnic communities (Karlsson, 2006). As western societies become more materialistic, modernized, and alienating for individuals, it nevertheless remains likely that people of diasporic backgrounds will continue to seek ancestral pasts in the lands of their forebears, notably as host communities see the benefits in promoting personal heritage tourism.

References

Akhtar, S. (1999), "The Immigrant, the Exile, and the Experience of Nostalgia", *Journal of Applied Psychoanalytic Studies*, 1(2): 123-130.

Ali, A. and Holden, A. (2006), "Post-colonial Pakistani Mobilities: The Embodiment of the 'Myth of Return' in Tourism", *Mobilities*, 1(2): 217-242.

Ancestral Journeys of Scotland (2007), "Ancestral Journeys of Scotland", Available <http://www.ancestraljourneysofscotland.com/main.htm> (Accessed January 20, 2007).

Ancestral Roots Travel (2007), "Bring Your Past into the Present: Welcome to Ancestral Roots Travel", Available <http://www.ancestralrootstravel.com/> (Accessed April 5, 2007).

Anthias, F. (1998), "Evaluating 'diaspora': beyond ethnicity?", *Sociology*, 32: 557-580.

Ashworth, G.J. (1995), "Heritage, Tourism and Europe: A European Future for a European Past", in Herbert, D.T. (ed.), *Heritage, Tourism and Society*, London: Mansell, pp. 68-84.

Asiedu, A. (2005), "Some Benefits of Migrants' Return Visits to Ghana", *Population, Space and Place*, 11: 1-11.

Bartlett, T. (2001), "Virginia Develops African-American Tourism Sites", *Travel Weekly*, 4 June: 16.

Basu, P. (2001), "Hunting Down Home: Reflections on Homeland and the Search for Identity in the Scottish Diaspora", in Bender, B. and Winer, M. (eds), *Contested Landscapes: Movement, Exile and Place*, Oxford: Berg, pp. 333-348.

Basu, P. (2004a), "My Own Island Home: The Orkney Homecoming", *Journal of Material Culture*, 9(1): 27-42.

Basu, P. (2004b), "Route Metaphors of 'Roots-Tourism' in the Scottish Highland Diaspora", in Coleman, S. and Eade, J. (eds), *Reframing Pilgrimage: Cultures in Motion*, London: Routledge, pp. 150-174.

Basu, P. (2005a). "Macpherson Country: Genealogical Identities, Spatial Histories and the Scottish Diasporic Clanscape", *Cultural Geographies*, 12: 123-150.

Basu, P. (2005b), "Pilgrims to the Far Country: North American 'Roots-Tourists' in the Scottish Highlands and Islands", in Ray, C. (ed.), *Transatlantic Scots*, Tuscaloosa: University of Alabama Press, pp. 286-317.

Belk, R.W. (1990), "The Role of Possessions in Constructing and Maintaining a Sense of Past", in Goldberg, M.E., Gorn, G. and Pollay, R.W. (eds), *Advances in Consumer Research*, vol. 17, Provo, Utah: Association for Consumer Research, pp. 669-676.

Bhatia, S. and Ram, A. (2001), "Rethinking 'acculturation' in relation to diasporic cultures and postcolonial identities", *Human Development*, 44: 1-18.

Birthright Israel (2007), "Taglit, Birthright Israel: Your Adventure, Your Birthright, Our Gift", <www.birthrightisrael.com> accessed January 5, 2007.

Bly, L. (1998), "Keeping Up with the Joneses", *USA Today*, 7 August: D1-D2.

Bradish, C. and Bradish, P. (2000), "Doing Genealogy on the Road", *Everton's Genealogical Helper*, 54(2): 44.

Bruner, E.M. (1996), "Tourism in Ghana: The Representation of Slavery and the Return of the Black Diaspora", *American Anthropologist*, 98(2): 290-304.

Bunce, M. (1994), *The Countryside Ideal: Anglo-American Images of Landscape*, London: Routledge.

Bushman, C.L. (1980), "Joy in Our Heritage", in Kimball, C.E. (ed.), *Joy*, Salt Lake City: Deseret Book, pp. 95-107.

Card, D., DiNardo, J. and Estes, E. (1999), "The more things change: immigrants and the children of immigrants in the 1940s, the 1970s and the 1990s", in Borjas, G.J. (ed.), *Issues in the Economics of Immigration*, Chicago: University of Chicago Press, pp. 227-269.

Carter, S. (2004), "Mobilizing *Hrvatsko*: Tourism and Politics in the Croatian Diaspora", in Coles, T. and Timothy, D.J. (eds), *Tourism, Diasporas and Space*, London: Routledge, pp. 188-201.

Clarke, K.M. (2004), *Mapping Your Networks: Power and Agency in the Making of Transnational Communities*, Durham: Duke University Press.

Cohen, E.H. (2006), "Religious Tourism as an Educational Experience", in Timothy, D.J. and Olsen, D.H. (eds), *Tourism, Religion and Spiritual Journeys*, London: Routledge, pp. 78-93.

Cohen, R. (1997), *Global Diasporas*, London: Routledge.

Coles, T. and Timothy, D.J. (2004), "'My Field is the World': Conceptualizing Diasporas, Travel and Tourism", in Coles, T. and Timothy, D.J. (eds), *Tourism, Diasporas and Space*, London: Routledge, pp. 1-29.

Compart, A. (1999), "Genealogy Travel: TW Writer's Road to Shanghai", *Travel Weekly*, 58(99): 1, 18.

Davis, F. (1979), *Yearning for Yesterday: A Sociology of Nostalgia*, New York: Free Press.

Dikomitis, L. (2004), "A Moving Field: Greek Cypriot Refugees Returning 'Home'", *Durham Anthropology Journal*, 12(1): 7-20.

Du, Y. (2006), "Translocal Lineage and the Romance of Homeland Attachment: the Pans of Suzhou in Qing, China", *Late Imperial China*, 27(1): 31-65.

Dunne, J.G. (1989), "Looking for Mr. Burns", *Condé Nast Traveler*, 24(8): 118-129.

Duval, D.T. (2002), "The Return Visit-Return Migration Connection", in Hall, C.M. and Williams, A.M. (eds), *Tourism and Migration: New Relationships between Production and Consumption*, Dortrecht: Kluwer, pp. 257-276.

Essah, P. (2001), "Slavery, Heritage and Tourism in Ghana", *International Journal of Hospitality and Tourism Administration*, 2(3/4): 31-49.

European Focus (2007), "Private Genealogy Research and Discovery Tours", Available <http://www.eurofocus.com/html/genealogy_tours.html> (accessed April 5, 2007).

Fein, J. (2006), "Tapping the Power of Your Ancestors", *Spirituality and Health*, July: 36-39.

Feng, K. and Page, S.J. (2000), "An Exploratory Study of the Tourism, Migration-Immigration Nexus: Travel Experiences of Chinese Residents in New Zealand", *Current Issues in Tourism*, 3(3): 246-281.

Ferguson, J. (1995), "Salsa and Lederhosen", *Geographical*, 67(2): 22-24.

Friedmann, J. (1999), "The Hybridization of Roots and the Abhorrence of the Bush. In Featherstone, M. and Lash, S. (eds), *Spaces of Culture: City-Nation-World*, London: Sage, pp. 13-39.

Gade, D.W. (1994), "Germanic Towns in Southern Brazil: Ethnicity and Change", *Focus*, 44(1): 1-6.

Genealogy Germany (2007), "Your Personal Heritage Tour", Available <http://www.genealogy-germany.de/index.htm> (accessed April 5, 2007).

Gordon, E.T. and Anderson, M. (1999), "The African diaspora: toward an ethnography of diasporic identification", *Journal of American Folklore*, 112: 282-296.

Graham, B., Ashworth, G.J. and Tunbridge, J.E. (2000), *A Geography of Heritage: Power, Culture and Economy*, London: Arnold.

Gupta, A. and Ferguson, J. (1992), "Beyond 'culture': space, identity and the politics of difference", *Cultural Anthropology*, 7(1): 6-23.

Hall, C.M. (2007), "Response to Yeoman et al.: The Fakery of 'the Authentic Tourist'", *Tourism Management*, 28: 1139-1140.

Hannam, K. (2004), "India and the Ambivalences of Diaspora Tourism", in Coles, T. and Timothy, D.J. (eds), *Tourism, Diasporas and Space*, London: Routledge, pp. 246-260.

Hasty, J. (2002), "Rites of Passage, Routes of Redemption: Emancipation Tourism and the Wealth of Culture", *Africa Today*, 49(3): 47-76.

Hayes, B.J. (1997), "Claiming our Heritage is a Booming Industry", *American Visions*, 12(5): 43-48.

Hewison, R. (1987), *The Heritage Industry: Britain in a Climate of Decline*, London: Methuen.

Hollinshead, K. (1998), "Tourism and the Restless Peoples: A Dialectical Inspection of Bhabha's Halfway Populations", *Tourism, Culture and Communication*, 1(1): 49-77.

Hooked on Genealogy Tours (2007), "Genealogical Jaunts, 2008 & 2007", <http://www.hookedongenealogytours.com/> (accessed April 5, 2007).

Horst, H.A. (2004) "A Pilgrimage Home: Tombs, Burial and Belonging in Jamaica", *Journal of Material Culture*, 9(1): 11-26.

Hou, F. and Balakrishnan, T.R. (1996), "The Integration of Visible Minorities in Contemporary Canadian Society", *Canadian Journal of Sociology*, 21(3): 307-326.

Hudman, L.E. and Jackson, R.H. (1992), "Mormon Pilgrimage and Tourism", *Annals of Tourism Research*, 19: 107-121.

Hungarian National Tourist Office (2000), *Hungary: Routes to Your Roots*, Budapest: Hungarian National Tourist Office.

Jobe, M.S. (2004), "Roots Tourism in The Gambia and the Development of Heritage Sites around Jufure", in UNESCO (ed.), *Africa 2009: Conservation of Immovable Cultural Heritage in Sub-Saharan Africa, Report on the 5th Regional Thematic Seminar on Sustainable Tourism and Immovable Culture*, Paris: UNESCO World Heritage Center, pp. 33-39.

Jorgensen, L.W. (1988), "A 'Roots' Vacation", *Ensign*, 18(6): 23.

Karlsson, L. (2006), "The Diary Weblog and the Travelling Tales of Diasporic Tourists", *Journal of Intercultural Studies* 27(3): 299-312.

Keating, S.K. (1990), "Genealogy Business Isn't Blarney", *Insight*, 6(22): 46-47.
Kelly, M.E. (2000), "Ethnic Pilgrimages: People of Lithuanian Descent in Lithuania", *Sociological Spectrum*, 20(1): 65-91.
Kemp, R. (2000), "Appointment in Ghana: An African American Woman Unravels the Mystery of Her Ancestors", *Modern Maturity*, July-August: 1-17.
Keng, K.A. (1997), "Assessing Macro Environment Trends in Singapore: Implications for Tourism Marketers", *Asia Pacific Journal of Tourism Research*, 1(2): 5-14.
Kurzweil, A. (1995), "Genealogy as a Spiritual Pilgrimage", *Avotaynu*, 11(3): 16-20.
Lambert, R.D. (1996), "The Family Historian and Temporal Orientations towards the Ancestral Past", *Time and Society*, 5(2): 115-143.
Lambert, R.D. (2006), "Descriptive, Narrative, and Experiential Pathways to Symbolic Ancestors", *Mortality*, 11(4): 317-335.
Legrand, C. (2005), "Nation, Migration and Identities in Late Twentieth-Century Ireland", *Narodna umjetnost—Hrvatski časopis za etnologiju i folkloristiku*, 41(1): 18-29.
Lew, A.A. and Wong, A. (2004), "Sojourners, *guanxi* and Clan Associations: Social Capital and Overseas Chinese Tourism to China", in Coles, T. and Timothy, D.J. (eds), *Tourism, Diasporas and Space*, London: Routledge, pp. 202-214.
Louder, D.R. (1989), "Le Québec et la Franco-Américanie: A Mother Country in the Making", in Hornsby, S., Konrad, V. and Herlan, J. (eds), *Four Hundred Years of Borderland Interaction in the Northeast*, Fredericton, NB: Acadiensis Press, pp. 126-136.
Louie, A. (2003), "When You are Related to the 'Other': (Re)locating the Chinese Homeland in Asian American Politics Through Cultural Tourism", *Positions*, 11(3): 735-763.
Lowenthal, D. (1996), *Possessed by the Past: The Heritage Crusade and the Spoils of History*, New York: Free Press.
Lowenthal, D. (1979), "Environmental Perception: Preserving the Past", *Progress in Human Geography*, 3: 549-559.
Lowenthal, D. (1975), "Past Time, Present Place: Landscape and Memory", *Geographical Review*, 65(1): 1-36.
Lynch, K. (1972), *What Time Is This Place?* Cambridge, MA: MIT Press.
Magenheim, H. (2002), "The Romanian Connection", *Travel Weekly*, 24 June: 92.
Matthiessen, P. (1989), "The Captain's Trail", *Condé Nast Traveler*, 24(1): 106-134.
Meethan, K. (2004), "'To Stand in the Shoes of My Ancestors': Tourism and Genealogy", in Coles, T. and Timothy, D.J. (eds), *Tourism, Diasporas and Space*, London: Routledge, pp. 139-150.
Mitchell, K. (1997), "Different Diasporas and the Hype of Hybridity", *Environment and Planning D: Society and Space*, 15: 533-553.
Morgan, N. and Pritchard, A. (2004), "Mae'n Bryd I ddod Adref—It's Time to Come Home: Exploring the Contested Emotional Geographies of Wales", in Coles, T. and Timothy, D.J. (eds), *Tourism, Diasporas and Space*, London: Routledge, pp. 233-245.

Nash, C. (2002), "Genealogical Identities", *Environment and Planning D: Society and Space*, 20: 27-52.

Nauck, B. (2001), "Intercultural contact and intergenerational transmission in immigrant families", *Journal of Cross-Cultural Psychology*, 32(2): 159-173.

Neville, G.K. (2003), *Kinship and Pilgrimage: Rituals of Reunion in American Protestant Culture*, working paper 22, Atlanta: Emory Center for Myth and Ritual in American Life.

Nguyen, T.H. and King, B. (2004), "The Culture of Tourism in the Diaspora: The Case of the Vietnamese Community in Australia", in Coles, T. and Timothy, D.J. (eds), *Tourism, Diasporas and Space*, London: Routledge, pp. 172-187.

Olsen, D.H. (2006), "Tourism and Informal Pilgrimage Among the Latter-day Saints", in Timothy, D. and Olsen, D.H. (eds), *Tourism, Religion and Spiritual Journeys*, pp. 254-270. London: Routledge, pp. 254-270.

Parker, J.C. (1989), *Going to Salt Lake City to Do Family History Research*, Turlock, CA: Marietta Publishing.

Richards, S.L. (2005), "What is to Be Remembered?: Tourism to Ghana's Slave Castle-Dungeons", *Theatre Journal*, 57: 617-637.

Rodriguez, D.H. (1987), "More than Names", *Ensign*, 17(1): 12.

Safran, W. (1991), "Diasporas in Modern Societies: Myth of Homeland and Return", *Diaspora*, 1(1): 83-99.

Schramm, K. (2004), "Coming Home to the Motherland: Pilgrimage Tourism in Ghana", in Coleman, S. and Eade, J. (eds), *Reframing Pilgrimage: Cultures in Motion*, London: Routledge, pp. 133-149.

Scottish Genealogy Tours (2003—copyright), "Time to Come Home", Available <http://www.scottishgenealogytours.co.uk/index1.htm> (accessed December 15, 2006)

Select Travel Service (2007), "Welcome to Select Travel Service: Europe's Bespoke Genealogical Travel Service", Available <http://www.selectgenealogytours.com/> (accessed April 5, 2007).

Sheffer, G. (1986), "A New Field of Study: Modern Diasporas in International Politics", in Sheffer, G. (ed.), *Modern Diasporas in International Politics*, London: Croom Helm, pp. 1-15.

Smith, G. (1999), "Transnational Politics and the Politics of the Russian Diaspora", *Ethnic and Racial Studies*, 22(3): 500-523.

Soja, E. (1996), *Thirdspace: Journeys to Los Angeles and Other Real-and-Imagined Places*, Cambridge, MA: Blackwell.

Stephenson, M.L. (2002), "Travelling to the Ancestral Homelands: The Aspirations and Experiences of a UK Caribbean Community", *Current Issues in Tourism*, 5(5): 378-425.

Sutton, C.R. (2004), "Celebrating Ourselves: The Family Reunion Rituals of African-Caribbean Transnational Families", *Global Networks*, 4(3): 243-257.

Thorstenson, C.T. (1981), "Discovering my Icelanders", *Ensign*, 11(8): 25.

Timothy, D.J. (1997), "Tourism and the Personal Heritage Experience", *Annals of Tourism Research*, 24(3): 751-754.

Timothy, D.J. (2001), "Genealogy, Religion and Tourism", Paper presented at the Association of American Geographers Annual Conference, New York City, 1-3 March.
Timothy, D.J. (2002), "Tourism and the Growth of Urban Ethnic Islands", in Hall, C.M. and Williams, A.M. (eds), *Tourism and Migration: New Relationships between Production and Consumption*, Dortrecht: Kluwer, pp. 135-151.
Timothy, D.J. and Boyd, S.W. (2003), *Heritage Tourism*, London: Prentice Hall.
Timothy, D.J. and Boyd, S.W. (2006), "Heritage Tourism in the 21st Century: Valued Traditions and New Perspectives", *Journal of Heritage Tourism*, 1(1): 1-17.
Timothy, D.J. and Teye, V.B. (2004), "American Children of the African Diaspora: Journeys to the Motherland", in Coles, T. and Timothy, D.J. (eds), *Tourism, Diasporas and Space*, London: Routledge, pp. 111-123.
Toronto Ukrainian Genealogical Group (2007), "Discover Your Roots Tour to Western Ukraine", Available at www.torugg.org/TUGG%20Projects/trip_to_ukraine.html (accessed January 20, 2007).
Tremblay, R. (2006), *Floribec: Espace et Communauté*, Ottawa: University of Ottawa Press.
Tuan, Y.F. (1974), *Topophilia: A Study of Environmental Perception, Attitudes and Values*, Englewood Cliffs, NJ: Prentice Hall.
Wales Tourist Board (n.d.), *Tracing Your Ancestors*, Cardiff: Wales Tourist Board.
Webster, C. and Timothy, D.J. (2006), "Travelling to the 'Other Side': The Occupied Zone and Greek Cypriot Views of Crossing the Green Line", *Tourism Geographies*, 8(2): 162-181.
Woodtor, D.P. (1993), "African-American Genealogy: A Personal Search for the Past", *American Visions*, 8(6): 20-23.

Chapter 8

Genealogy as Religious Ritual: The Doctrine and Practice of Family History in the Church of Jesus Christ of Latter-day Saints

Samuel M. Otterstrom

Introduction

People of diasporic, hybrid, or unknown ancestry share a growing fascination in understanding their genetic and historical roots, notably roots that have been transplanted to places far from their forebears' homelands. As other chapters in this volume indicate, this increasing interest typically involves a desire to carry out genealogical research as a way for people to understand where they "are from". Among the millions of people undertaking family history research, most may be classified as either serious or occasional amateur genealogists, with the majority conducting genealogy and family history as one of several part-time hobbies.

Many people with a desire to seek their roots soon discover that the Church of Jesus Christ of Latter-day Saints (also more popularly known as the Mormon Church or LDS Church—in this chapter both the full name of the church and "Church" are used interchangeably) is a serious advocate for genealogy. Its members have a deep interest in family history, perhaps more than any other group in the world, with many belonging to the category of "serious genealogists" in a religious sense that is actually poorly explained by both the conventional amateur and professional genealogist secular categories. What is the explanation for this peculiar religious emphasis and how does it relate to the religious teachings and practices of the Mormon Church? This chapter outlines the history and doctrinal background of genealogy in the LDS Church. It also underscores the part the Church has played in the growth of genealogical research as a religious imperative for its members. In the process of encouraging genealogy for doctrinal reasons, the Church has developed resources for the millions of people outside of Mormonism who enjoy genealogy and family history as a hobby or profession.

Before explaining the history and background of Mormons' interest in genealogy I will place Mormon religious practices in context within two cultural constructs. These are first, activities performed to fulfill religious or cultural obligations and second, the emergence of personal heritage and identity seeking in connection with genealogical practices. The Mormon faith's emphasis on genealogy in the context of specific ritual practices translates into a particular type of heritage-seeking developed through religious obligation, which begets a distinct pattern of personal heritage behaviors, including travel that focuses on places and people of the past (see Timothy, 1997). This chapter is aimed primarily at the non-Mormon reader, as it illustrates the religious foundations of family history and genealogy among Latter-day Saints and highlights the doctrines and practices that drive the practice today.

Religious Obligation and the Personal Heritage Quest

Religions of the world each have their own sets of requirements of action and belief that adherents are expected to follow. For example, in many Christian denominations baptism and partaking of the sacrament of communion are required. In Islam, five key pillars of belief, including the frequency and manner of prayer, almsgiving, and pilgrimage to Mecca, form the foundations of Muslim religious doctrines. In the Church of Jesus Christ of Latter-day Saints members are required not only to participate in certain religious ordinances such as baptism for themselves, but they are also asked to share their beliefs with non-Mormons and pursue genealogy in order to identify ancestors who were not Church members, often because these ancestors lived before the founding of the Mormon faith in 1830. The call to share their beliefs has translated into the Church's large proselytizing program that includes more than 50,000 missionaries around the world (Deseret News, 2005). The second admonition for Mormons, to seek out their ancestry, has been intimately tied to both the growth of the Church's genealogical program and to the multiplication of the religion's most revered houses of worship: its temples (Durrant, 1983; Oaks, 1989).

Catherine Nash (2002) linked genealogical research to an expanded concept of personal identity. Her research into the cultural practices of tourists involved in Irish genealogy pursuits reveals the evident connection between learning the identity of one's ancestors and the questions that can arise concerning the nationalities or cultural status of that personal identity for people with multiple ethnic backgrounds.

Personal identity is shaped and influenced in part at least, as a person discovers his/her ancestral roots, and consequently a related desire or urge often emerges, which focuses on the concept of heritage, of re-creating, preserving, and identifying with past environments in which one's ancestors lived. David Lowenthal's (1996, 1985, 1975) extensive work on developing the theoretical construct of heritage underscores how modern society values such attention to places that can exist only as attempted re-creations or memories of past places. In terms of genealogical practices, individuals' efforts to connect with the people and places of their ancestral pasts can be classified as personal heritage activities (Timothy, 1997). Family historians searching for records of their ancestors often participate in personal heritage tourism as a means to obtain primary data on the life and background of relatives in certain locales but also to experience the environments most closely associated with these relations (Meethan, 2004; Timothy, 1997).

Although many are aware of the emphasis that the LDS Church places on genealogical research, and serious genealogists may travel to search in the Church's archived records, far fewer understand why the Mormons are so involved in this activity. The academic literature has only scattered references to the importance of genealogical research within the theology of the Church. Some scholars mention the concept of religious obligation in their explanation of Mormon practices, but often it is discussed in passing while other subjects are emphasized such as genealogy's use in teaching history (Johnston, 1978; Parker, 1990), the potential of the Church's Family History Library as a historical research resource (Gerlach and Nicholls, 1975), or the structure and distinctive qualities of Mormon families (Anderson, 1937; Clydesdale, 1997; Wilkinson and Tanner, 1980). Davies (2000) does relate some of the distinctive Mormon religious practices such as genealogical research to their theological roots in more detail. Outside of writings published by authorities of the Church or its members, however, there is a need for more in-depth explanation of the religious reasons behind Mormon genealogical pursuits in a manner that is more accessible to non-Mormon readers. Additionally, this chapter provides a crucial example of how religious obligations can translate into personal heritage seeking and identity creation.

Mormon Temples, Genealogy, and the Religious Imperative

Visitors to the headquarters of the Church of Jesus Christ of Latter-day Saints in downtown Salt Lake City, Utah, will find an interesting array of office and religious buildings in a multi-block area known as Temple Square. The 28-story Church Office Building is where the day to day administrative operations of the Church are coordinated. It also houses archives of church-related documents, some of which are open to non-LDS scholars. The Tabernacle (home of the Mormon Tabernacle Choir) with its distinctive architecture, the historic Assembly Hall, and the expansive new Conference Center, where 21,000 people can meet in one hall (Dietsch, 2002), all serve as Church meeting and cultural event venues. Nearby is the Joseph Smith Memorial Building, which was named after the Church's first prophet, who lived between 1805 and 1844. It was transformed from its former use as a hotel to its current multiple function capacity, having restaurants, reception areas, Church office and meeting space, and a movie theater used to present Church-produced films to visiting audiences. This facility also has areas where visitors can search for their ancestors on computers as they are assisted by Church missionaries, who are assigned to help the public search for their ancestors on the computer system and answer questions concerning the Church's beliefs about genealogy. This is a good starting place for budding genealogists, but most family researchers seek out the nearby Family History Library. Its proximity to the most central buildings of the Mormon faith and the designation of missionaries to assist visitors with locating records of their ancestors are emblematic of the centrality of genealogy and family history to the Church's core sense of identity and its profession of outreach to the public.

The key to understanding why members of the Church of Jesus Christ of Latter-day Saints do family history research is not in the library itself, however, but in the historic temple located within Temple Square just east of the library. Mormon genealogical interests are directly linked to temple worship. The Salt Lake Temple is the focal point of the whole Church Headquarters complex and is probably one of the best-known icons of "Mormonism" (Oman and Snyder, 1997).

Today the Salt Lake City temple is just one of 124 such edifices operated by the Church around the world, and additional temples are planned as Church membership expands. These buildings are different from LDS chapels, or churches, where regular Sunday services are held. Indeed, temples are reserved for non-Sabbath worship and sacred rites. Because the temple is the most sacred space among Mormons, only members of the Church who hold a "temple recommend", or a certificate which verifies their observance of the Church's standards of belief and practice, may enter the temple (Packer, 1999: 21).

In the temple, Latter-day Saints perform sacred ordinances or rites on behalf of their deceased family members and ancestors who were unable to perform these ordinances for themselves while living. These ordinances carried out by the living on behalf of the dead include baptisms and sealings—that is, rituals wherein families, including ancestors, are "sealed" or bonded together for eternity (Nelson, 2002; Sorensen, 2003). The ceremonies also include the "endowment", which is comprised of instruction to participants about the creation of the earth, the atonement and sacrifice of Jesus Christ, God's eternal plan for His children, and covenants to obey the laws and commandments of God (Packer, 1980; Sorensen, 2003; Talmage, 1912). All of these rites are considered essential ordinances for all of humankind, whether living or dead, for salvation, and can only be carried out for the dead once adequate information about each individual is known (e.g. name, birthdate, parents' names)—hence, the need for genealogical data. Whereas baptisms of the living can be performed in regular church baptismal fonts and in natural bodies of water such as rivers, baptisms for the dead can only be done in a temple.

Latter-day Saints' claim of authority to baptize both the living and on behalf of the dead stems back to Joseph Smith, the founder and first president of the Church, who established the new faith in upstate New York in 1830 (Black and Telford, 2004; Porter and Black, 1988). He and Oliver Cowdery, another early leader, testified that they had received authority (in the form of a religious ordination, which Mormons term "receiving the priesthood") to baptize from the angel John the Baptist on May 15, 1829 (*Doctrine and Covenants* 1986, section 13).[1] Later these men received the power to confer the gift of the Holy Ghost (to have the constant companionship of the Holy Spirit—a gift given to all newly baptized Church members) and to perform other church ordinances by the angels Peter, James, and John (Hinckley, 1979: 21-23).[2] Based on these experiences and the belief that other Christian faiths lack proper authority, the LDS Church asserts that it is the faith with the same power that John the Baptist himself possessed when he baptized Jesus Christ. Furthermore, Latter-day Saints subscribe to the edict found in John 3:5 in the New Testament, which reads: "Unless a man be born of water and the spirit he cannot enter the

---

1   Priesthood (McConkie, 1966: 69-72) is the power and authority given by God and Jesus Christ to administer the affairs of the Church. Mormons practice baptism by complete immersion in water by one who has the priesthood authority to baptize.

2   In LDS theology, angels are God's messengers, who may be either spirits or physical beings who have died and been resurrected into tangible bodies or translated as were Enoch and Elijah. John the Baptist, Peter, James and John are referred to as the resurrected form of angels (*Bible Dictionary*, 1983).

kingdom of God".[3] Mormons believe that this injunction includes the living, as well as the deceased who have not had the opportunity to receive the essential baptism in mortality. Consequently, devout Mormons are baptized on behalf of their dead, often for many people over many years as genealogical research reveals the names of additional ancestors.

Like much of Christianity, Latter-day Saints believe that baptism by one with proper authority and by full immersion of the physical body is essential for salvation. Baptism is a cleansing ritual, through which individuals are washed clean from their sins and commit to obey the will of God. This is a central belief, for "the unrighteous shall not inherit the kingdom of God" (1 Corinthians 5: 9) and "no unclean thing can dwell with God" (*Book of Mormon*, 1986 [1 Nephi 10: 21]). Because the deceased are spirit beings, they are unable to participate in this requirement. Therefore, it must be done in the flesh on behalf of those who were not baptized during their mortal existence by someone who has already completed this requirement for himself (McConkie, 1966; Nelson, 2002).

Among various scriptural passages in support of the practice of baptism for the dead are the words of Paul in 1 Corinthians 15: 29, "For what shall they do who are baptized for the dead if the dead rise not at all? Why are they then baptized for the dead?" In this context, Paul is teaching about the doctrine of the resurrection and is suggesting that the ordinance of baptisms for the dead would be meaningless without the resurrection brought about by the death and resurrection of Christ.

Furthermore the Latter-day Saints connect that verse to 1 Peter 4: 6, where Peter describes the role of Jesus Christ in initiating the preaching of His gospel to those beyond the veil of death: "For this cause was the gospel preached also to them that are dead, that they might be judged according to men in the flesh, but live according to God in the spirit." This second verse supports the LDS belief that there are persons waiting for baptism and other sacrosanct ordinances to be performed in their behalf because they have accepted the message of the Savior in the spirit world, or the place where spirits await the resurrection.

Temple marriages performed by proxy on behalf of the dead are called "sealings". Accordingly, ancestral couples are sealed together by the proper authority and thereby receive the opportunity to be together in the afterlife in married companionship. Additionally, parent to child sealings are also performed with the belief that extended family relationships will also continue in the hereafter. Marriage sealings for the deceased are only done for those who were formally married in their mortal lives, while children who have died are sealed to their natural or adopted parents.

Mormons believe that as individuals continue to exist in the spirit world after death, they will have a say in the matter of proxy work done for them. Gordon B. Hinckley, current President of the Church explained:

> In the spirit world these same individuals are then free to accept or reject those earthly ordinances performed for them, including baptism, marriage, and the sealing of family relationships. There's no compulsion in the work of the Lord, but there must be opportunity. (Hinckley, 1999: 17)

---

3   All Bible verses are from the King James Version.

Mormons further believe that while they should perform these ordinances first for themselves, and then for the dead, they should especially seek out their personal forebears in order to do the proxy baptisms, endowments, and sealings that are essential for eternal life. Joseph Fielding Smith, tenth president of the Church, underscored this point further:

> [I]t is the man's duty to go to the temple, have his wife sealed to him and have their children sealed, so that the family group, that unit to which he belongs, is made intact so that it will continue throughout all eternity. That is the first duty that a man owes to himself, to his wife, and to his children...Then it is his duty to seek his record as far back as he can go and do the same thing for each unit. He should begin with his father and mother and their children, and his grandfather and his children, great-grandfather and his children, and have the work done in like manner, linking each generation with the one that goes before. That is the responsibility resting upon every man who is at the head of a household in this Church. (Smith, 1955: 206-207)

This belief is based largely on the prophecy of the Old Testament prophet Malachi (Malachi 4: 5-6): "Behold, I will send you Elijah the prophet before the coming of the great and dreadful day of the Lord: And he shall turn the heart of the fathers to the children, and the heart of the children to the fathers, lest I come and smite the earth with a curse". Referring to these verses and the role of Elijah, the prophet Joseph Smith explained in 1842, "It is sufficient to know, in this case, that the earth will be smitten with a curse, unless there is a welding link of some kind or other, between the fathers and the children, upon some subject or other—and behold what is that subject? It is the baptism of the dead. For we without them cannot be made perfect; neither can they without us be made perfect" (*Doctrine and Covenants*,1986, section 128:18). The doctrine is that the central social unit in the eternities is the family (Hinckley, 1999; Nelson, 2006; Smith, 1955).

In sum, there is a two-step process that members of the Church of Jesus Christ of Latter-day Saints are asked to complete. First is to be baptized themselves at the age of accountability, deemed to be eight years of age (or later if the situation warrants, such as conversion from another faith), and then as adults to receive their endowment and be married and sealed to their spouse in one of the temples.[4] The second step is for members to seek vital information about their deceased ancestors (e.g. their names, birth dates and places, and parents' names) and then perform baptisms, endowments, and sealings in proxy for these relatives.[5] Much of the temple work is done or overseen by people who are called or set apart as missionaries or "temple workers" for this purpose.

---

4 The LDS Church does not believe in "original sin" and therefore does not practice infant baptism; it has established a policy of eight years as the youngest one can be baptized and the time when people know right from wrong. Any child who dies before turning eight is not baptized by proxy, and is believed to be saved by virtue of his or her innocence of sin and the mercy and grace of Jesus Christ. For a doctrinal explanation of this belief see Moroni chapter 8 in *The Book of Mormon* (1986).

5 Mormons can also undertake baptisms, endowments, and sealings for the deceased who were not related to them.

Historical Background of Temple Ordinances

Joseph Smith organized the Church of Jesus Christ of Latter-day Saints in 1830 in Palmyra, New York. Soon, church headquarters were moved to Kirtland, Ohio. The first temple of the Church was built there and dedicated in March 1836. The Kirtland Temple was different from later temples in that it was not a place for ordinances for the dead, but it was seen as a holy structure where Joseph Smith and other leaders received heavenly visitations, gave instruction, and pronounced blessings and instruction to the living (see Talmage, 1912: 121-123; Brown, 1999: 209-217). Significantly, Joseph Smith and Oliver Cowdery testified that Malachi's prophesied return of Elijah occurred in the Kirtland Temple (*Doctrine and Covenants*, 1986, section 110: 13-16), where Elijah delivered the sealing authority to Joseph Smith (Packer, 1980: 131-142). The Latter-day Saints were only in Kirtland for a short while longer, as they were led by Joseph Smith to settle in Missouri and then Illinois. Temples were planned for Mormon settlements in Independence and Far West, Missouri, but they were never built because the Mormons were forcibly expelled from the state by mobs who disagreed with several of their practices and beliefs (Baugh, 2001; Smith, 1979). The next center of the LDS faith was Nauvoo, Illinois, beginning in 1839. This settlement on the Mississippi River grew very quickly in the years that followed.

Just over a year after the Mormons began occupying Nauvoo, the practice of baptizing for the dead was announced by Joseph Smith in 1840 on the occasion of a funeral for one of the prophet's bodyguards. Soon members of the Church were being baptized on behalf of their deceased ancestors and friends, in places such as the Mississippi River. At first, records were not kept and many of the baptisms were performed with little oversight. After October 3, 1841, baptisms for the deceased were stopped while a baptismal font was constructed in the basement of the unfinished Nauvoo Temple. The new font was used from November 21, 1841, to January 1845, even as some baptisms for the dead were administered again in the Mississippi River because the Nauvoo Temple font was very busy. By January 1845, some 15,722 baptisms for the dead had been recorded (Black and Black, 2002: i-vi).

Religious practices in the completed Nauvoo Temple extended beyond baptisms for the dead, but for just a very short period. Endowments and some sealings were performed for the living in the temple beginning in December 1845 and continued through the first months of 1846 (Talmage, 1912: 133; Anderson and Bergera, 2005). Joseph Smith was killed by a mob on June 27, 1844, but the Latter-day Saints followed the instructions he had previously given to other leaders of the Church regarding the manner in which the endowment should be administered.

The practice of performing eternal ordinances for the dead, which began with baptisms in Nauvoo, was resumed in the Latter-day Saints' new home in Utah. While temples were being built in Salt Lake City, St. George, Logan, and Manti (all in Utah), a temporary structure known as the Endowment House was constructed and utilized in Salt Lake City. The Endowment House was used for baptisms, endowments and sealings from 1855 to 1889, but mostly for the living (Cowan, 1989: 67-73; Ludlow, 1992: 456).

Endowments for the dead were first performed in the St. George Temple, in southern Utah, which was the first completed Utah temple (1877) (Smith 1955, 171). The St. George Temple was followed by one in Logan (1884) in the northern part of the territory, Manti in central Utah (1888), and then the well-known Salt Lake Temple (1893). The Salt Lake Temple remains the largest of all the temples of the Church (*Deseret News*, 2005: 565). With their completion, baptisms for the dead, endowments and sealings for the living and the dead were all now being performed in these newly constructed edifices.

Prior to 1894, temple work on behalf of the dead was often performed in a disorganized manner, and there was not clear direction concerning the practice of sealings of children to parents (Bennett, 1960: 101-102). On April 8, 1894 the President of the Church, Wilford Woodruff announced:

> We want the Latter-day Saints from this time to trace their genealogies as far as they can and to be sealed to their fathers and mothers. Have children sealed to their parents, and run this chain through as far as you can get it...This is the will of the Lord to his people. (Woodruff, 1922: 149)

A 1960 Church Sunday School book further teaches:

> When sealings for a family group are performed in a temple, there is being welded one shining link in a chain of celestial pattern. When each parent is sealed, with brothers and sisters to his parents, two additional links are forged. This process must be continued on back to the beginning, family group by family group, till the chain of...authority is complete with no missing link. (Bennett, 1960: 196)

Development of the Genealogical Society of Utah

Only a few months after Wilford Woodruff's announcement, the Genealogical Society of Utah was organized in the fall of 1894 to help Latter-day Saints seek out their own genealogies (Allen et al., 1995). The first president of the society was Franklin D. Richards, one of the twelve apostles of the Church, which demonstrates the Church's interest in the new association.[6] The society library grew slowly, having only some 2,000 volumes of genealogies by 1911. At first, the library was open to paying members, but in 1944 the Church incorporated it into its administrative structure, and the library's holdings became available to the public (Ludlow, 1992; Allen et al., 1995). Church members were encouraged to submit their family group sheets (standard format pages listing single nuclear families with their vital information) to the library. Consequently, being even distantly related to members of the LDS Church can be an advantage to a non-Mormon family historian, as the Mormon relatives may have already contributed an extensive pedigree of shared ancestors. In subsequent years, further growth at the library resulted in holdings

---

6 The highest governing body in the Church is the "Quorum of the First Presidency", which consists of the president or prophet of the Church and two counselors. The next level of organization is the "Quorum of the Twelve Apostles", which has twelve members as its name implies (Lds.org, 2007).

of some 61,754 volumes and 2,050 manuscripts by 1960 (Bennett, 1960). These holdings included government manuscript census records, copies of parish registers, lists of ships' passengers, published pedigrees, and the like.

Only two other American genealogical societies predated the Genealogical Society of Utah. They were the New England Historic and Genealogical Society (est. 1844) and the New York Genealogical and Biographical Society (est. 1869). Additionally, a year after the Genealogical Society of Utah's inception, the Genealogical Society of Pennsylvania was organized in 1895. In the twentieth century numerous other societies were started in the United States to promote genealogical research. The Utah society, because of its religious orientation, has sought to collect records from all over the United States and the world, however, in contrast to most other genealogical societies that are more localized in their focus (Scott, 1969), or are restricted to specific ethnic groups, or to people linked to particular historical events, such as Mayflower descendants.

Since 1960, the growth of the Church's genealogical library has been significant. It is now known as the Family History Library and has over 742,000 microfiche, 310,000 books, 4,500 periodicals, and 2.4 million rolls of microfilm, with some 4,100 rolls of microfilm and 700 books and 16 electronic resources being added to the collection every month in 2003. Additionally, about 1,900 patrons use the library daily, many of whom are not members of the Church of Jesus Christ of Latter-day Saints (Familysearch.org, 2006; Warren and Warren, 2001). Some of the library's workers are paid employees, but the majority is made up of volunteer missionaries who assist people in finding their roots. Many library staff members are fluent in foreign languages, which helps foreign patrons and English speakers who need to search foreign records.

In Salt Lake City, the Family and Church History Department at Church headquarters has been charged with helping people around the world learn more about their ancestors. The Department's work has extended to many nations as it works to gather names of the world's deceased. For example, for many years the Church has sent employees and missionaries throughout the world to microfilm genealogical records. The efforts of these microfilmers have extended to over 110 countries, territories, and possessions, and the records they have collected have formed the basis of the extensive records available in the Family History Library. There are currently some 200 microfilm cameras being used in over 45 different countries (FamilySearch.org, 2006).

The Church has also developed a website that provides free database access to names and basic data of hundreds of millions of names of people from the past. FamilySearch.org was launched on May 24, 1999, and has grown rapidly since that time (Rasmussen, 2000). One distinct database available at the site is the International Genealogical Index (IGI). The IGI is essentially a complete database of names (with dates, places, and family relationships) of people who have died and who were members of the Church or deceased individuals whose temple ordinances (baptism, endowment, and sealings) have been performed by proxy in one of the temples. FamilySearch.org now has over one billion names that can be searched within its databases (FamilySearch.org, 2006).

The Mormons are very serious about preserving genealogical documents. The original 2.4 million rolls of microfilm, which contain millions of names on reproduced censuses, probate records, church records, etc., are stored in the Granite Mountain Records Vault, which is located in Little Cottonwood Canyon in the Wasatch Mountains southeast of Salt Lake City. It was constructed between 1960 and 1966 by blasting crisscrossing tunnels into the canyon face, and as such was designed to protect and preserve the Church's valuable genealogical records against possible man-made and natural disasters (Allen et al., 1995). The Church is now in the process of digitizing all of these microfilms using specialized computer technology, with the end goal of making the records available on the Family Search website (www.familysearch.org) (Hart, 2006).

Besides the main Family History Library in Salt Lake City, the Church has established some 4,500 branch family history libraries around the world (Hart, 2006). Most of these libraries are located in churches and are staffed by volunteers from the local Church membership. They are open to Latter-day Saints and the general public, and patrons are assisted by librarians in both doing basic and more advanced research. People can also order any of the microfilms from the main library in Salt Lake City and search them using microfilm readers in the local Church family history library. Currently, some 100,000 rolls of microfilm from Salt Lake circulate each month (FamilySearch.org. 2006).

One issue with the genealogical work done by Church members over the years is duplication. This has resulted in temple rites, such as baptism, being done multiple times for certain of the deceased. Recognizing this situation, Nephi Anderson, a former secretary of the Genealogical Society of Utah predicted in October 1911, that "...as temples multiply, and the work enlarges to its ultimate proportions, this Society [Genealogical Society of Utah] or some organization growing out this society, will have in its care some elaborate, but perfect system of exact registration and checking, so that the work in the temples may be conducted without confusion or duplication..." (Anderson, 1912: 22).

Many efforts have been developed since that time to rectify the duplication situation. The establishment of the Temple Index Bureau (TIB) in 1922 was the first major step (Bennett, 1960: 174; Allen et al., 1995). Workers in the bureau were tasked to arrange cards alphabetically with the names of all those who had temple ordinances completed; this amounted to some 22,000,000 cards in 1960 (Bennett, 1960: 103). The International Genealogical Index has incorporated and now replaced the TIB, as its records are now all available online through FamilySearch.org. Members are asked to compare their family records with the IGI before they submit names of their predecessors to temples for ordinance work. These indexing efforts have resulted in a tremendous database of vital records that are valuable to Church members and the public alike.

Even with these developments, some duplication remains. Gordon B. Hinckley, current Church President, announced in 2005 the development of a system to end this challenge. He said, "We...have been engaged for some time in a very difficult undertaking. To avoid such duplication, the solution lies in complex technology. Preliminary indications are that it will work, and if this is so, it will be a truly remarkable thing with worldwide implications" (Hinckley, 2005: 3). The new system,

"Family Tree", is still being developed, but its product manager, Jim Greene, noted that Family Tree is the first step toward ending duplication of temple ordinances and, hopefully, maybe even some research" (Lloyd, 2006: 5).

In sum, the collection, preservation, and dissemination of this vast amount of genealogical data have been made possible through the direct administrative influence and financial support of the Church of Jesus Christ of Latter-day Saints for the clear purpose of enabling Mormons and others to pursue their personal genealogical research. The religious aims of the Church in connection with temple ordinances must also, of necessity, be remembered in explaining this remarkable emphasis on genealogical research.

Personal and Family History, Religious Obligation, and Personal Heritage

Not only does the Church of Jesus Christ of Latter-day Saints emphasize the importance of documenting ancestral relationships and performing proxy temple ordinances, it also encourages recordkeeping at a more personal and immediate level (Kimball, 2003; Wells and Wells, 1986). Members of the Church have the charge to write down their own history as well as that of their families. Ezra Taft Benson, Church President from 1985 to 1994, said in 1978, "Our responsibility to keep a journal and to write our own personal histories and those of our ancestors, particularly those who belong to the first four generations of our pedigree, has not changed" (Benson, 1978: 30). According to Spencer W. Kimball (2003: 32), a recent Church President,

> ...we urge our young people to begin today to write and keep records of all the important things in their own lives and also the lives of their antecedents in the event that their parents should fail to record all the important incidents in their own lives. Your own private journal should record the way you face up to challenges that beset you. Do not suppose life changes so much that your experiences will not be interesting to your posterity. Experiences of work, relations with people, and an awareness of the rightness and wrongness of actions will always be relevant...No one is commonplace, and I doubt if you can ever read a biography from which you cannot learn something from the difficulties overcome and the struggles made to succeed. These are the measuring rods for the progress of humanity...Your journal is your autobiography, so it should be kept carefully. You are unique, and there may be incidents in your experience that are more noble and praiseworthy in their way than those recorded in any other life. There may be a flash of illumination here and a story of faithfulness there...What could you do better for your children and your children's children than to record the story of your life, your triumphs over adversity, your recovery after a fall, your progress when all seemed black, your rejoicing when you had finally achieved? Some of what you write may be humdrum dates and places, but there will also be rich passages that will be quoted by your posterity.

The Church admonition on journal writing and the creation of personal histories is intertwined with each member's responsibility to do genealogy and temple work. Boyd K. Packer (1980: 229-230), one of the Church's twelve apostles, outlined the following list of "genealogical and temple work" duties:

1. Each of us is to compile his or her own life history.

2. Each of us is to keep a Book of Remembrance (a record of personal and family legacies, stories, and photographs).

3. As individuals and families we are each to seek out our kindred dead, beginning first with the four most recent generations on each line, and then going back as far as we can.

4. We are each to participate in other programs such as name extraction [transcribing of census and other vital records] when asked to do so.

5. We are to organize our families and hold meetings and reunions.

6. If we have access to a temple, each of us should go to the temple as often as possible to do ordinance work.

Church members are thus counseled not only to do genealogy research so that temple ordinances can be performed for their deceased ancestors, including adopted and illegitimate children, but they also are to develop strong family relationships and identities through reading, writing, and sharing the life histories of their ancestors, living relatives, and themselves. This process produces a distinct personal heritage component in Latter-day Saint lives and families that is seen as both obligatory and personally beneficial (see Durrant, 1981).

The culture of record keeping in the Church goes back to its early beginnings with the early presidents, such as Wilford Woodruff writing in his journal daily for thirty-five years, and Joseph Smith recording many of his experiences even while he led the fledgling religious organization through its turbulent early years (Cannon, 1977: 5). The call to "organize our families" has helped multiply family organizations by Church members. Although not an official part of the Church organization, or limited to Latter-day Saints, Church-inspired family reunions are a prevalent part of Mormon culture. Their purpose is to connect relatives together and provide a venue for disseminating information on common ancestors.

In essence, the work of writing journals, holding family gatherings, researching the vital events and life histories of ancestors, and completing temple ordinances all center on the same core LDS doctrine related to their belief in the afterlife—the doctrine of an eternal and united family. Church President Joseph Fielding Smith stated it simply: "We believe that the family will go on" (Smith, 1955: 67). Because of this belief in the continuance of families beyond the grave, it follows that people would want to become more familiar with their progenitors beyond doing temple ordinances on their behalf. Furthermore, writing personal histories preserves the memories of the living, which can be passed down to children and grandchildren so these descendants can know the characters and lives of their grandmothers and grandfathers. All of these facets taken together underscore how the religious obligation of doing family history and temple work translates into a powerful personal heritage experience for the Mormon faithful.

References

Allen, J.B., Embry, J.L. and Mehr, K.B. (1995), *Hearts Turned to the Fathers: A History of the Genealogical Society of Utah, 1894-1994*, Provo, UT: BYU Studies.

Anderson, D.S. and Bergera, G.J. (eds) (2005), *The Nauvoo Endowment Companies, 1845-1846: A Documentary History*, Salt Lake City: Signature Books.

Anderson, N. (1937), "The Mormon Family", *American Sociological Review*, 2(5): 601-608.

Anderson, N. (1912), "Genealogy's Place in the Plan of Salvation", *Utah Genealogical and Historical Magazine*, 3:12-22.

Baugh, A.L. (2001), "From High Hopes to Despair: The Missouri Period, 1831-39", *Ensign*, 31(7): 44.

Bennett, A.F. (1960), *Saviors on Mount Zion*, Salt Lake City: Deseret News Press.

Benson, E.T. (1978), "Worthy of All Acceptation", *Ensign*, 8(11): 30.

Bible Dictionary (1983 printing), Salt Lake City: The Church of Jesus Christ of Latter-day Saints.

Black, S.E. and Black, H.B. (2002), *Annotated Record of Baptisms for the Dead 1840-1845: Nauvoo, Hancock County Illinois* (Vol. 1), Provo, UT: Brigham Young University Press.

Black, S.E. and Telford, J. (2004), *Joseph Smith: Praise to the Man*, Orem, UT: Millennial Press.

Book of Mormon (1986 printing), Salt Lake City: Church of Jesus Christ of Latter-day Saints.

Brown, M.B. (1999), *The Gate of Heaven: Insights on the Doctrines and Symbols of the Temple*, American Fork, UT: Covenant Communications, Inc.

Cannon, E. (1977), *Putting Life in Your Life Story*, Salt Lake City: Deseret Book.

Clydesdale, T.T. (1997), "Family Behaviors among Elderly U.S. Baby Boomers: Exploring the Effects of Religion and Income Change, 1965-1982", *Social Forces*, 76(2): 605-635.

Cowan, R.O. (1989), *Temples to Dot the Earth*, Salt Lake City: Bookcraft.

Davies, D.J. (2000), *The Mormon Culture of Salvation*, Aldershot, UK: Ashgate.

Deseret News (2005), *2006 Church Almanac*, Salt Lake City: Deseret News.

Dietsch, D. (2002), *Building the Church of Jesus Christ of Latter-day Saints Conference Center*, New York: Edizioni Press.

Doctrine and Covenants (1986 printing), Salt Lake City: Church of Jesus Christ of Latter-day Saints.

Durrant, G. (1981), "The Importance of Personal and Family History", In *World Conference on Records: Preserving our Heritage, August 12-15, 1980*, vol. 1, series 2, Salt Lake City: Church of Jesus Christ of Latter-day Saints.

Durrant, G. (1983), "Genealogy and Temple Work: You Can't Have One Without the Other", *Ensign*, 13(8):18.

Familysearch.org (2006), "The Largest Collection of Free Family History, Family Tree and Genealogy Records in the World", <http://www.familysearch.org/> (home page), accessed 13 October 2006.

Gerlach, L. and Nicholls, M. (1975), "The Mormon Genealogical Society and Research Opportunities in American History", *The William and Mary Quarterly*, 32(4): 625-629.

Hart, J.L. (2006), "Digitizing Hastens at Microfilm Vault", *Deseret News Church News*, 11 March: 3, 7.

Hinckley, G.B. (1979), *Truth Restored*, Salt Lake City: Deseret Book.

Hinckley, G.B. (1999), "Why These Temples?", In Church of Jesus Christ of Latter-day Saints (ed.), *Temples of the Church of Jesus Christ of Latter-day Saints*, Salt Lake City: Church of Jesus Christ of Latter-day Saints, pp.14-19.

Hinckley, G.B. (2005), *Official Report of the One Hundred Seventy-fifth Semiannual General Conference of the Church of Jesus Christ of Latter-day Saints*, Church of Jesus Christ of Latter-day Saints: Salt Lake City, pp. 2-4.

Johnston, A.M. (1978), "Genealogy: An Approach to History", *The History Teacher*, 11(2): 193-200.

Kimball, S.W. (2003), "New Era Classic: The Angels May Quote from It", *New Era*, February: 32. (reprinted from 1975 original).

Lds.org (2007), "Gospel library, Gospel topics, Church administration", <http://www.lds.org/> (home page), accessed 26 April 2007.

Lloyd, R.S. (2006), "Looking Ahead: Tantalizing Changes in View for Family History Work", *Deseret News Church News*, 25 March: 5.

Lowenthal, D. (1975), "Past Time, Present Place: Landscape and Memory". *Geographical Review*, 65: 1-36.

Lowenthal, D. (1985), *The Past is a Foreign Country*, Cambridge: Cambridge University Press.

Lowenthal, D. (1996), *Possessed by the Past: The Heritage Crusade and the Spoils of History*, New York: Free Press.

Ludlow, D.H. (ed.) (1992), *Encyclopedia of Mormonism*, New York: Macmillan.

McConkie, B.R. (1966), *Mormon Doctrine* 2nd edition, Salt Lake City: Bookcraft.

Meethan, K. (2004), "To Stand in the Shoes of My Ancestors: Tourism and Genealogy", in Coles, T. and Timothy, D.J. (eds), *Tourism, Diasporas and Space*, London: Routledge, pp. 139-150.

Nash, C. (2002), "Genealogical Identities", *Environment and Planning D: Society and Space* 20: 27-52.

Nelson, R.M. (2002), "Prepare for Blessings of the Temple", *Ensign*, 32(3): 17.

Nelson, R.M. (2006), "Nurturing Marriage", *Ensign*, 36(5): 36-38.

Oaks, D.H. (1989), "Family History: 'In Wisdom and in Order'", *Ensign*, 19(6): 6.

Oman, R.G. and Snyder, J.P. (1997), "Exterior Symbolism of the Salt Lake Temple: Reflecting the Faith that Called the Place into Being", *BYU Studies*, 36(4): 6-68.

Packer, B.K. (1980), *The Holy Temple*, Salt Lake City: Bookcraft.

Packer, B.K. (1999), "The Holy Temple." in Church of Jesus Christ of Latter-day Saints (ed.), *Temples of the Church of Jesus Christ of Latter-day Saints*, Salt Lake City: Church of Jesus Christ of Latter-day Saints, pp. 20-27.

Parker, J.C. (1990), "Local History and Genealogy in the Classroom: Without Cooperation between Teachers and Librarians, Research Assignments Risk Failure", *The History Teacher*, 23(4): 375-381.

Porter, L.C. and Black, S.E. (1988), *The Prophet Joseph: Essays on the Life and Mission of Joseph Smith*, Salt Lake City: Deseret Book.

Rasmussen, R.C. (2000), "Computers and the Internet in the Church", in Sperry Symposium (ed.), *Out of Obscurity: The LDS Church in the Twentieth Century*, Salt Lake City: Deseret Book, pp. 274-285.

Scott, K. (1969), "Genealogical Societies: Introduction, General Description and Brief History", in *World Conference on Records and Genealogical Seminar: Papers*. [Area K: Genealogical Societies: Hereditary and Lineage Societies], Salt Lake City: Genealogical Society of the Church of Jesus Christ of Latter-day Saints.

Smith, J.F. (1955), *Doctrines of Salvation, Volume 2*, Salt Lake City: Bookcraft.

Smith, J.F. (1979), *Essentials in Church History*, Salt Lake City: Deseret Book.

Sorensen, D.E. (2003), "The Doctrine of Temple Work", *Ensign*, 33(10): 56-63.

Talmage, J.E. (1912), *The House of the Lord: A Study of Holy Sanctuaries Ancient and Modern*, Salt Lake City: Deseret News Press.

Timothy, D.J. (1997), "Tourism and the Personal Heritage Experience", *Annals of Tourism Research*, 34: 751-754.

Warren, P.S. and Warren, J.W. (2001), *Your Guide to the Family History Library*, Cincinnati: Betterway Books.

Wells, G. and Wells, G.J. (1986), "Hidden Benefits of Keeping a History", *Ensign*, 26(7): 47.

Wilkinson, M.L. and Tanner, W.C. (1980), "The Influence of Family Size, Interaction, and Religiosity on Family Affection in a Mormon Sample", *Journal of Marriage and the Family*, 42(2): 297-303.

Woodruff, W. (1922), "The Law of Adoption", *Utah Genealogical and Historical Magazine*, 13: 145-1

Chapter 9

Genetics, Genealogy, and Geography

David C. Mountain and Jeanne Kay Guelke

This paper is a critical review of family historians' use of DNA test results, as a set of cultural practices with several prominent geographical components. We specifically concentrate on the leisure pursuit of family history, rather than on the discipline of genetics and genome projects in cultural context (discussed by Rabinow, 1999 and Greenhough, 2006); but discuss scientific findings to the degree that they inform genetic (or "deep") genealogy, an avocation demonstrating a strong appetite for accurate scientific data. (See, for example, the popular overviews of genome studies by Wells, 2006; Sykes, 2006; and Oppenheimer, 2006.) In response to Greenhough and Roe's (2006) concern with "how people are actually making sense of the possibilities posed by biotechnological innovation," we consider how DNA research encourages family historians to rethink basic geographic concepts like ethnicity, migration, diffusion, and spatial correlation.

In acknowledging Nash's (2004, 2006) and Simpson's (2001) concern that genetic genealogy may only reinforce ethnic stereotyping in locales that already exhibit inter-ethnic tension, we argue that in practice, the incorporation of genetics into family history is equally likely to subvert traditional beliefs about ethnicity and nationality. Genetic genealogy can be read in multiple ways, notably depending upon family historians' own beliefs about their roots and whether they claim membership within an ethnically-defined state, a diasporic community, an assimilationist civic nation (Ignatieff, 1994), or a trans-national culture. Moreover, the diversity of stakeholders within the community of genetic genealogists and the democratizing effects of cyberspace provide for lively round-tables of debate and opportunities to refute bigoted or essentialist positions. This is not to posit genetic genealogy as a model of bias-free impartiality or logic, as examples of ethnocentrism are easy enough to find; but it is to argue that the meanings of DNA test results to consumers and scientists are in a current state of dispute and change, and are likely to remain so for some time.

As with many aspects of popular culture in modern societies, "genetic genealogy" or "deep genealogy" functions extensively through the Internet. Our evidence consequently derives principally from qualitative interpretations of American-based Internet sources as today's principal public forum among the "communities of practice" (Wenger, 2001; Dodge and Kitchin, 2001; Simpson, 2000) of genetic genealogists.

Strathern (1995), Simpson (2000), Palsson (2002), Brodwin (2002), Nelkin and Lindee (2004), and Nash (2004, 2005, 2006) extensively critiqued genome projects and genetic genealogy, and cautioned of their potential to reinforce old concepts of patriarchy, racism, and pseudo-scientific criteria for belonging to or exclusion from a specific territory. Their concerns over "the popular appropriation of genetics"—in which scientists are also embedded (Nelkin and Lindee, 2004: 5)—are particularly meaningful for contested places like Northern Ireland, where publicity over new data on a territory's "original" and presumably rightful inhabitants, can have political consequences.

Santos and Maio (2004: 367-77) in their study of a genome project in Brazil, however, warned against assuming that the racialized politics and debates of one context or nation necessarily apply to other places, where social circumstances might be very different. Tutton (2002: 116) interviewed Orkney Islands residents about their involvement in a highly publicized genetic project, and found that most participated out of intense interest in their family history, notably those with family lore about descent from shipwrecked Spanish sailors. "Rather than looking to have some essential Orcadian identity confirmed, some participants talked of genealogies that would place them in relation to other parts of the world....many also saw that to be Orcadian could mean to be a composite of many different elements." Fleising (2001) and Skinner (2006) noted that some of the concerns over "genetic essentialism" unnecessarily ignored the ambiguity and anti-racist tenor of many DNA test results and studies.

The Role of Genetics in Genealogy

Genetic genealogy, defined here as the use of DNA test results to reveal information about a donor's ancestry, has several foundations. A main one is the scholarly scientific study of the DNA of various human populations for purposes ranging from curiosity-driven research about human origins and migrations to concerns for curing genetically-caused diseases (Cavalli-Sforza, Menozzi, and Piazza, 1994). This scholarly genetic research is not to be confused with genetic genealogy, but genealogists and DNA testing firms are avid consumers of this knowledge base.

Family historians' searches for ancestors often reach a point (known in genealogy circles as "the brick wall") when they cannot locate more archival records. Some turn to the alternative method of DNA testing in an effort to retrieve their personal heritage. Family historians who are deeply assimilated within a hybridized society like the United States may not know where their "Old World" ancestors originated, and hope that DNA testing will reveal genetic affinities to particular societies and places. Particularly difficult for African American family historians is the search for records that link them to particular cultures and regions of Africa. Some Latino genealogists in the U.S. see DNA testing as a way to develop links back to the conquistadores or to specific indigenous nations. Some whites hope to confirm or refute family legends that they have "Native blood" (Smolyenak and Turner, 2004; Pomery, 2004; Shriver and Kittles, 2004; Winston and Kittles, 2005). DNA testing may offer a way to extend their searches beyond the earliest written records and family recollections.

Family historians today are able to go beyond written records and memories of older relatives through DNA testing by commercial companies: Family Tree DNA (http://www.familytreedna.com/), Oxford Ancestors (www.oxfordancestors.com), Ethnoancestry (www.ethnoancestry.com), African Ancestry (www.africanancestry.com), DNA Heritage (www.dnaheritage.com), and Relative Genetics (www.relativegenetics.com) are among the prominent firms. Using techniques long known to forensic and medical geneticists (and paternity-testing companies), these for-profit firms will analyze small samples of clients' Y-DNA (males), mitochondrial DNA, or autosomal DNA (for males or females) that are collected through cheek swabs. These companies then perform specified tests on the samples and send clients print-outs of their results, with fees charged according to the number and types of analyses ordered.

On a voluntary basis, clients may agree to let their DNA testing company keep their samples for possible further analysis; or to be put in touch via e-mail with other clients who share similar genetic patterns, in the hope that people with close genetic matches can determine their common ancestry (Smolenyak, 2004). (See for example, the website "ysearch" of Family Tree DNA at <http://www.ysearch.org/>.) If two unacquainted men showed the same results for 26 out of 26 tested markers, for example, they would likely be related through a common ancestor who lived within the past few centuries.

Most analyses performed to date involve men testing for Y-DNA genetic markers, which are passed down strictly from father-to-son; or both men and women testing for mitochondrial DNA (mtDNA) markers, which pass through the maternal line. Obviously peoples' genetic inheritance is a complex mix of both male and female lines. Some tests consequently involve autosomal DNA, which offers a more composite picture of the sum of one's ancestry, but their results to date are limited to the likelihood that a client has a given percentage of ancestry from generalized, often continental-scale groups like Africans or Native Americans.

The Y-DNA markers come in two forms: single nucleotide polymorphisms (SNPs) and short tandem repeats (STRs). The SNP markers mutate very slowly, about one chance in 500 million per location per generation (Vallone, 2003); while the STR markers mutate much more rapidly, about one chance in 500 per location per generation (Heyer et al., 1997). The mutation rate for mtDNA markers in the so-called "hypervariable regions" is about one chance per 30,000 per location per generation, slower than for Y-DNA STRs but faster than for Y-DNA SNPs.

The Y-DNA tests have consequently generated the most interest among genealogists because STR markers appear to mutate more rapidly than the other types, thus better permitting analysis within an historic time frame (Underhill, Shen, et al., 2000; but see Helgason et al., 2003). In the case of a small number of ethnic groups (defined below), Y-DNA markers may suggest a man's ancestral affinity through his paternal line, or even his descent from a legendary patriarch of the Middle Ages (Moore, et al., 2006; Zerjal, Xue, et al., 2003).

Because DNA test results are so far limited to a few kinds of readings and a comparatively small sample base of clients, they consequently are more likely to result in merely a print-out of genetic markers for the puzzled novice (termed a "newbie" by more experienced genetic genealogists), rather than to extensions of the family tree or even an ethnic identity. Because mtDNA mutates very slowly, test results may take a woman's maternal line directly back to the Pleistocene, with no information on the recent millennia, let alone centuries (Richards et al., 1996).

Commercial Y-DNA testing has accordingly stimulated a large number of "surname projects" on the Internet as well as some scholarly surname studies (Sykes and Irven, 2000; Jobling, 2001; Pomery, 2004) to which men voluntarily contribute their Y-DNA test results. These projects operate on the principle that both surnames and y-DNA markers are passed down through the paternal line. Theoretically, unacquainted men with the same surname should show similar or even identical genetic patterns (haplotypes) if they shared a common paternal ancestor. If a close match is found, the donors are probably related, and may even dismantle one another's "brick walls" by sharing their accumulated archival genealogical information. Surname projects are particularly popular among men of Scottish and Irish ancestry who carry the names of clans or septs, respectively, that supposedly derive from an eponymous male ancestor and a specific territory (Hill, Jobling, and Brady, 2000a, 2000b).

Much of the discussion today among deep genealogists (vs. professional geneticists, physical anthropologists, or molecular biologists) occurs today on Internet message boards and forums such as the "DNA: GENEALOGY-DNA Mailing List" sponsored by RootsWeb.com (http://lists.rootsweb.com/index/other/DNA/GENEALOGY-DNA.html) and the American International Society of Genetic Genealogy (http://www.isogg.org/). Unfortunately, the genetic genealogy companies' proprietary restrictions on use of the hundreds of intriguing e-mail messages and discussion-strings posted on their websites prohibits duplication of them for purposes other than private family history work; and the difficulties of acquiring participants' informed consent for use of their messages argue against quoting from them here. In a general way one can observe that the message boards reveal active, even passionate virtual "communities of practice" (Wenger, 2001) of deep genealogists. Many of the e-mail correspondents appear to be self-taught in interpreting DNA test results, archaeological evidence, and ancient history. Much of their discussion concerns how to interpret specific DNA markers in culturally meaningful terms.

Geographical Bases of Genetic Genealogy

Geography informs deep genealogy in several important ways. Geneticists learned early on in their analyses of human DNA that most major genetic groupings of donors (termed haplogroups) were few in number, and that they tended to cluster geographically such that it made sense to map their distributions, based upon the donors' or their earliest-known paternal or maternal ancestor's place of birth (Cavalli-Sforza, Menozzi, and Piazza, 1994). Genetic markers are believed to be the most common in the population at their place of origin (e. g. mutation) and will often exhibit gradients or clines with distance from this site, although long-distance migrations of founder populations to newly colonized areas make this principle imprecise in application.

Most geographical distributions of genetic markers examined to date do not coincide well with textbook ethnic group distributions and tend to operate on wider, even continental scales (Shriver and Kittles, 2004; Rootsi, Magri, et al., 2004). Such discrepancies may reflect the different criteria used by societies who define themselves as ethnic groups: "primordialists" depend upon a common ancestry defined in a genealogical sense; but others (such as the English) define themselves according to a set of shared culture traits and history, and may acknowledge multiple founding populations and shifting locations according to different historical periods (Weinreich, Bacova and Rougier, 2003; Romanucci-Ross, De Vos and Tsuda, 2006). The search for a hereditary basis to ethnic affinities nevertheless continues. It sometimes exhibits reasonably unequivocal results in the case of culturally distinctive or isolated groups like the Saami (Lapps) (Rosenberg, Pritchard, et al., 2002); but exhibits ambiguity or diverse founder populations for others.

Scholarly DNA-testing of populations is often conducted on a national or regional basis, such as genome surveys of Portugal or Sicily (Renfrew and Boyle, 2000). Although modern political boundaries may not coincide well with haplogroup distributions, genetic research is often funded by national scientific funding agencies, whose focus within their borders may be influenced by nationalist politics. Primordialist beliefs about nationality based upon common ancestry may motivate some studies—and the media interpretations of their findings—as scientists are hardly isolated from their socio-political milieus. However, geneticists work from a variety of different motives, including the desire to understand and cure genetically based disorders such as Tay-Sachs disease and sickle-cell anemia, illnesses whose distributions are unaffected by national boundaries.

Today the spatial coverage of DNA test sites is incomplete, and the selection and number of locales from which DNA is sampled can bias efforts to portray national or regional profiles. Data collection sites are heavily skewed toward Western Europe (Helgason et al., 2000) despite the efforts of the Human Genome Diversity Project <http://www.stanford.edu/group/morrinst/hgdp.html?> and other research teams to generate comprehensive genetic overviews.

The National Geographical Society (NGS) currently attempts to correct this spatial imbalance through its Genographic Project (https://www5.nationalgeographic.com/genographic/atlas.html) in partnership with IBM's Computational Biology Center and the commercial firm of Family Tree DNA (http://www.familytreedna.com/ftdna_genographic.html). This project involves both worldwide DNA sampling of small indigenous populations by scholars, as well as voluntary DNA donations from anyone willing to submit and pay for the processing of a cheek swab sample for inclusion in the NGS data base. The Genographic Project offers donors the opportunity to learn about their own DNA test results, and thereby to search for distant relatives (http://www.isogg.org/gss.htm), as well as to contribute altruistically to the larger scientific objectives. A major goal of the NGS's genetic testing of large and diverse population samples is to trace the global prehistory of human migration patterns (Wells 2006). Charges of commercial genetic "mining" of indigenous people plagued earlier genome projects, and NGS promises that a portion of its profits will contribute to cultural preservation for Native groups. (https://www5.nationalgeographic.com/genographic/atlas.html).

DNA testing does have the potential to illuminate questions of past population migrations and cultural diffusion, and at a range of temporal and spatial scales. For example, did the Anglo-Saxon "invasion" of England occur as a mass-migration (which would be expected to show up in the Y-DNA of a large percentage of today's Englishmen) or did Anglo-Saxon culture traits principally diffuse across England, leaving its earlier gene pool basically unaltered? Did the Viking migrations consist mostly of male warriors, or did family groups move together? Scholars currently disagree, but some use DNA test results to promote their arguments (Sykes 2006; Wilson, Weiss, et al., 2001; Weale, Weiss, et al., 2002; Oppenheimer, 2006; McEvoy and Edwards, 2005).

Where DNA surname projects fail to yield meaningful results, some deep genealogists now turn to "geographic projects" or "heritage projects" defined by the earliest-known locale of one's ancestors (such as East Anglia) or by a unique ethnic heritage (such as Flemish or Roma), respectively. Participants hope to locate relatives who do not share the same surname, and who consequently cannot be reached through surname projects. Others hope to unravel the genetic secrets of a region or nation from which they claim descent (http://www.genetealogy.com/resources/html/cat7.html or http://www.familytreedna.com/surname.asp).

Genetic Identities in Place and Time

The geographic distributions of genetic markers are consequently of great interest to geneticists, historians, and social scientists; as well as to family historians interested in locating their pedigrees within a specific social history context. The further hope among some family historians is that a simple cheek swab will reveal an "Irish gene" or an "Ibo gene" that will confer some roots in a particular ethnic group and homeland, or perhaps even a demonstrable tie to an extended kinship network through a surname project. Scientists', however, caution against facile interpretations of DNA data (Shriver and Kittles, 2004).

At issue for Simpson (2000) and Nash (2004, 2006) is how the intersection of deep-seated ethnic strife and media popularizations of genetic studies will play out in places like Britain and Ireland. We suggest that the current diversity of geographical backgrounds and political ideologies of deep genealogists will problematize the issue of racialized identities. Much voluntary DNA testing today is commissioned by Americans who have no strong sense of ethnic identity or involvement in the politics of a particular contested territory, due to assimilation and "melting pot" ethnic backgrounds. DNA testing for them becomes a search for a place-based personal identity rather than a means to exclude newcomers from a piece of turf. A personal identity quest via DNA testing may be open to criticism as a misplaced sense of insecurity, but its political implications seem less "dystopian" (Skinner, 2005) than critics of genetic essentialism (Simpson, 2000; Brodwin, 2002) have charged.

The rhetoric of American DNA testing companies' web sites often explicitly links their products to their clients' assumed sense of alienation and rootlessness, rather than to current territorial disputes. Identifying African homelands through DNA testing is particularly important to some descendants of African slaves, for whom early written records are non-existent. At the University of Massachusetts Lowell, The African-American DNA Roots Project uses "specific DNA analysis techniques to attempt to identify unique signature sequences among African-Americans that might link them to particular West African tribes" (http://www.uml.edu/dept/biology/rootsproject/africanamericandna.htm).

Genetic Genealogy advertises, "Learn more about your ancestors, where they came from, and their ethnic origins" (http://www.dnaancestryproject.com/ydna_intro_ancestry.php). A company named DNA Tribes asks, "What's your tribe?" and argues somewhat more moderately (http://www.dnatribes.com/faq.html) :

> Your top matches are the places in our database where your DNA profile is most common. A match with a particular ethnic or national population sample does not guarantee you or a recent ancestor (parent or grandparent, for instance) are a member of that ethnic group. However, a match does indicate a population where your combination of ancestry is common, which is most often due to shared ancestry with that population... Identity is complex and defined by a very personal combination history, geography, and culture. While a DNA test alone cannot define a person's identity, results from our analysis can provide important clues to ancestral origins within major world regions and sometimes even individual ethnic groups. Alongside genealogical, historical and cultural information, our analysis can contribute another important piece to the puzzle of personal identity.

While an ethnic identification can be difficult to deliver, depending upon the client's own ancestry and type of test ordered, they affirm that the quest for many clients is to claim a personal cultural identity rather than admission into the culture wars of a distant country; even if it does appear from their test results to be their ancestors' homeland.

The hope among both family historians and tourism bureaus is that someone with newly-found DNA markers that link him to a particular ethnicity or clan will actually travel to its ancestral sites as a tourist (Duffy, 2006); as currently occurs with conventional record-based genealogy tourism (Coles and Timothy, 2004). The concept of a genealogical homeland (defined here as a place of ancestral origins) operates at different scales, from the nation to a particular forebear's village, depending upon whether Y-DNA test results yield only general information, or link the donor to actual relatives and sites.

The ways in which these cultural practices can work in concert is illustrated by the following excerpt from a McCarthy family DNA-surname project web page. It suggests that putting the "pin in the map" is indeed central to their efforts.

> The McCarthy Surname Study will attempt through Y-DNA analysis to compare individuals with surnames common to the McCarthy clan for similarities and differences. It is hoped that distinct patterns will emerge that will allow McCarthy men to discover from which branch of the McCarthy tree they are descended. It is also hoped that these distinct patterns will eventually lead us to specific geographic areas of Ireland. For the many McCarthy males who do not know where their McCarthy ancestors lived in Ireland, this would be an exciting path for future research (McCarthy Surname Study, n. d.).

The Kerchner's Genetic Genealogy Y-DNA Surname Project (http://www.kerchner.com/kerchdna.htm) lists as one of its goals to find "ancestral home villages in Germany or a Germanic region of Europe" as an aid to further pursuit of genealogical records. An objective of the McDuffie family web site (http://www.mcduffiedna.

com/) is, "To determine the degree of relatedness among those with surnames of the great McDuffie diaspora and establish links with the ancestral homeland." Similarly, "The MacGregor DNA project exists to try to answer two questions - who are the MacGregors and where do they come from?" (http://www.clangregor.org/macgregor/dna.html) Although the search for a homeland is not a stated objective of many family surname websites, sometimes this is because the location of the earliest-known ancestors is already known and displayed; or because the commercial genetic genealogy forums themselves mount many family surname projects and provide the templates for project goals that lack an explicit homeland identity quest. At least the tacit (if not explicit) purpose of family surname-DNA web sites is to encourage contact among people with the same last name, who will ideally will be "blood" relatives—i. e., close genetic matches with other participants.

To be sure, prejudices may be found in these various DNA family and geographical projects; notably that prospective members must sometimes first establish their genetic identity or paternally-derived surname as rightfully belonging to a particular group. Adoptees apparently need not apply. On some Y-DNA sites, a "men only" sign hangs on the virtual door, even though women could otherwise participate by submitting test results from male relatives (and may surreptitiously do so.) Nevertheless these barriers appear permeable, as open-minded DNA genealogists appear to be more interested in pursuing their "true" roots as they understand them, than in reinforcing cultural stereotypes that poorly match the genetics. Moreover, a site that bars a potential member or community of interest can, in cyberspace, indirectly encourage another more inclusive web-based DNA project. The Internet-based DNA-surname and -geographical sites are today a far cry from the old elite American genealogical societies for descendants of colonial "first families" or of prominent statesmen.

Imagined Communities

Because both conventional and genetic genealogies engage with landscapes of memory, ethnicity, and nationality, this section first covers some relevant basic theses in critical cultural geography. This review underpins the subsequent discussion of the concepts of homeland and ethnicity in genetic genealogy discourse. We stress that the following categories are in practice dynamic, overlapping, and culturally situated, although specific groups may define themselves in fixed, essentialist ways.

Simpson (2001) noted the relevance of Benedict Anderson's (1991) thesis of "imagined communities" to genetic genealogy. Anderson observed that nations, homelands, and ethnic groups are to a large extent built upon stories of coherence, a shared history, and a common future. Nationalistic narratives are intended both to minimize differences within the same cohort across space, and also to convince people that they share a common identity with compatriots who predeceased them or who would be born in the distant future. Although actual evidence of such cultural unity may be lacking on the ground, the state or other large institutions like the church nevertheless have strong self-interests in solidifying their power through promoting a particular, unified vision of "one out of many" under their direction. *Hegemony*, or the normalizing of a single authoritative view of society that suppresses the many legitimate minority views, occurs when the state's official version of its people and events prevails voluntarily in public history, the media, and school curricula; and alternative versions are represented as unpopular or even deviant (Gramsci, 1971). Commercial interests are fully involved in this process, often as tourism bureaus or companies who stand to profit from branding places and ethnic cultures in particular ways (Edensor, 2002; Tunbridge and Ashworth 1996; Graham, Ashworth and Tunbridge, 2000).

Similarly, Cohen (1997: 26) characterized *diasporic communities* as sharing an idealized, even mythologized collective memory of the homeland, an ethnic or religious group identity, and a belief in a shared history and destiny. To be an active member of a diasporic community is to contemplate either a return migration to the homeland or at least to maintain personal ties with it through visits, cultural practices such as food customs and religious schools, and communications, political realities permitting (Coles and Timothy, 2004; Leeds-Hurwitz, 2005; Romanucci-Ross, De Vos, and Tsuda, 2006). As with the homeland population's own images of its identity, however, a true collectivity is logistically impossible because the group is too large and fragmented (Anderson, 1991). Its coherence can only be "imagined" through various narratives (such as history books) and practices (such as commemoration of a homeland's national holiday.)

Conversely, the experience of *assimilation* for immigrants' progeny—and oftentimes for people of indigenous descent—would mean no longer being able to push the "pin on a map" at the site of an ancestral community—or even to identify with an ethnic group—because its memory has been lost. Assimilation ultimately implies intermarriage over several generations, and adoption of the popular culture and allegiances that are normalized by the majority population within the host country; including its official stories and accepted practices of belonging and patriotism, rather than of the homelands of one's ancestors. Although the assimilation process could eventually confer a new ethnicity, yet most citizens of a post-settler society such as the United States know that their ancestral roots in its soil do not extend very far back in time, even if they do not know where they were.

Beyond these levels of territorial belonging are people whose family history of frequent movement or hybrid backgrounds—whether as children of peripatetic ancestors, of extensive intermarriage, or as members of today's high-tech business class in a globalizing economy—may give them little native-soil identification whatever. Feelings of alienation and displacement in a melting-pot, post-tribal world can nevertheless lead some individuals with placeless or "deterritorialized" identities to undertake genealogical research (Mitchell, 2000: 280; Friedman, 1994: 79; Meethan, 2004). They would necessarily approach genealogy without a clear understanding of their ancestors' homelands and ethnicities due to the dilution or disappearance of family ties with an ancestral locale or diasporic community. Genealogy then becomes an exercise in imagination, as much as of collecting evidence.

Mitchell (2000: 269) summarized the thesis of nationality as an imaginative process:

The notion that the community of the nation can only ever be imagined is important for understanding the ontological status of nations: it turns us away from the assumption that nations are somehow a *natural* result of a common people or folk and their relationship to a particular place, and towards the sense that nations are contested and thus a particular materialization of power and ideology.

Traditional memory- and archive-based family history follows Anderson's (1991) concept of imagined communities to the degree that family trees confer lists of ancestors, as well as historical roots in ethnicity or nationality, which can never be personally known. Simpson (2005) further postulated that "imagined genetic communities" rely upon the same impulses that generated nationalism in the past: capitalism and media (today, the Internet) capable of wide dissemination of information and ideas.

The concept of *identity* has a diffuse and complex scholarly and popular literature: different cultures define identity in different ways (Romanucci-Ross, De Vos and Tsuda, 2006). We use it here in the sense of Taylor and Spencer (2004: 1) that identity "embodies our sense of uniqueness as individual beings and as members of groups sharing values and beliefs" which, for purposes of this paper, include groups linked by values of kinship and ancestral origins. The practice of genealogy, whether archival or genetic, is clearly identity-related (Nash, 2002), as it confers a unique ancestry upon the individual; and, often, a sense of historically-based roots in an ethnic group or sub-culture and its territorial claims (Weinreich, Bacova and Rougier 2003). Particularly for the assimilated or transnational individual in a modern society, claiming an ancestral identity necessarily occurs more in the realm of thought and imagination than in a daily lived reality.

More to Mitchell's (2000: 269) point, above, regarding disputes over "particular materialization[s] of power and ideology," is that genetic genealogy has the potential either to reinforce or to undermine traditional definitions of homeland, ethnicity, kinship, and identity. Either way, it will no doubt develop and promote its own discourses of imagined communities, probably based in the agendas of geneticists who develop the methods and data, and of genealogists who "consume" it and use it to fashion their own narratives. Today their agendas overlap but are not identical. Genetic genealogy does include as its starting point conventional historical and cultural views of ethnicity, family, homeland, and nation; but because it incorporates new data sources and findings from scientists with no particular interest in family history pursuits or naturalized notions of nationality, it also has the potential to subvert biased assumptions about ancestry and ethnicity. Sometimes this displacement occurs through the non-scientific universalizing "family of Man" trope (Nash, 2004; 2005; 2006), but often the discourses are more locally based.

The Subversive Potential of Genetic Genealogy

One of the most highly publicized examples of the subversive potential of genetic genealogy in the United States concerned the longstanding belief among a small group of African Americans that they were descendants of a slave woman named Sally Hemings and Thomas Jefferson, the third president of the United States. DNA tests performed on both white and African American males claiming Jeffersonian paternity revealed that some—though not all—of these claims were plausible.[1] Within the context of the histories of racialized discrimination in the South and the Commonwealth of Virginia's lionizing of its notable statesmen; publication of the DNA test results was indeed explosive, notably for members of the whites-only Monticello Association of Jefferson's descendants (Williams, 2005).

Yet racialized and kinship identities and imagined communities figure in complex ways. One African American volunteer in the Hemings-Jefferson study (Williams, 2005) was apparently disappointed to learn that he was not a Jefferson descendant. For other African American males, learning that their Y-DNA links them back to a white slave-holder, rather than to an African ancestor, can be a deeply unsettling experience (Winston and Kittles, 2005: 221); just as some Scottish males must face the prospect that their Y-DNA does not match up with the majority of other members of their clan's surname project.

Genetic genealogy does not require the postulate that there are many different ethnic groups, in a fine-grained, ethnographic sense (such as Serbs and Croats,) or even idealized kinship norms of marital fidelity because it deals with genetic evidence, not culture traits. Indeed, the most common European haplogroups: the Y-DNA "R1b" group, and the mtDNA group "H", are widely distributed across Western Europe, comprising roughly 40% of today's populations. They are believed to have originated among hunter-gatherers in Pleistocene refugia of the Iberian Peninsula and southern France that sustained life beyond the ice cap and tundra that covered most of the continent. Although most common in these areas they are by no means restricted to them. Haplogroup distributions often show gradual gradients or clines across large distances and national boundaries.

Many of the more site-specific European haplotypes bear little relation to today's geographical patterns of ethnicity and nationality and have little to say about how *Blut und Boden* ("blood and soil") notions of ethnicity, at the scale of a small region or nation, come about. For example, despite the popular media inference of a pure and indigenous Irish haplotype in western Ireland (Nash, 2006); the actual geographical distributions of Hill, Jobling, and Bradley's (2000a, 2000b) "type 1" haplogroup indicate that both Ireland and Iberia are areas of heavy concentration (cf. Oppenheimer, 2006: 117), despite these two regions' few meaningful ethnic or political affinities in modern times. The Balkan states have comparable mixes of different haplogroups and share a high frequency of haplogroup I1b1*, despite their histories of inter-ethnic strife (Rootsi, Magri, et al., 2004).

At the very least, attempting to use geographically extensive DNA-defined groups to promote xenophobic ideologies of ethnicity and territory would reconfigure commonplace understandings of ancestrally-based ethnicity in ways that would dissatisfy those using genealogy to reinforce a sense of cultural belonging or nationalist affiliation in today's terms. There are some exceptions: a few groups with distinctive patterns of DNA, culture, and territory correlations reinforce rather than

---

1 The DNA tests would show only that the paternal ancestor carried a "Jefferson Y-chromosome." He could have been a near relative of Thomas Jefferson, although circumstantial historical evidence suggests that the president did father children with Sally Hemings.

undermine conventional views of ethnicity, such as the Saami (Lapps) and Sardinians. A few abrupt genetic differences between groups, such as sharp dissimilarity in the frequencies of haplogroup "I" between Latvians and Estonians, or Macedonians vs. Greeks, also support conventional models of ethnicity (Rootsi, Magri et al., 2004). Recent genetic findings, however, arguably have little to contribute to these groups' rhetoric of distinctive cultural identities that was not already in place.

Of some interest to people whose DNA markers do not differentiate at "downstream" times of more recent mutations, however, is the potential of DNA research to suggest whether their genetic ancestors could have arrived at their ancestral locale with the first waves of post-Pleistocene settlement. To Oppenheimer (2006), being genetically "English" probably means claiming descent from the original post-Pleistocene hunter-gatherers of the British Isles. He reads the genetic contributions of central European Celts, Romans, Anglo-Saxons, and Normans as slight, suggesting that hierarchical diffusion from a small body of politically powerful invaders or conquerors could explain English culture traits attributed to these groups.

Whatever light DNA research may shed on the "indigenous" people of Ireland or England, it is consequently unlikely to reinforce traditional nationalistic narratives of "Celtic Ireland" or of England as fundamentally "Anglo-Saxon" and rooted in rural agricultural village life. While Oppenheimer's (2006) conclusions could still be used by racists to exclude recent immigrants and their descendants from a sense of "belonging" to Ireland or to England, they further complicate the question of how one establishes the historical cut-off point for defining nationality, the basis upon which such a determination could be made, and just how seriously most citizens or governments are likely to take genetic testing in any event. The important issues of anti-immigrant sentiment in Ireland and Britain raised by Nash (2004, 2005, 2006) might be better addressed by protecting minorities' rights and enhancing their status as members of valued cultures, rather than by discouraging hobbyists or scientists who pursue genetic testing.

If one nevertheless assumes that ethnic groups and homelands are real enough, based upon current territorial conflicts and observable distinctive culture traits such as languages, then the genetic story is all about conditional probabilities. In other words, given a specific pattern of genetic markers, what is the probability that one's paternal (or maternal) ancestor belonged to the ethnic group of interest? In theory, a geneticist could then do detailed DNA analyses for all known ethnic groups and find the most likely group that matches an individual client's DNA pattern. This task might seem unnecessary to individuals whose ancestors apparently stayed rooted to the same spot for centuries, but it would enormously interest genealogists whose homeland memories have been lost, notably African-Americans (Timothy and Teye, 2004).

The problem currently is to get accurate estimates of the *a priori* probabilities based upon adequate sample sizes and systematic geographical and genetic coverage; and ideally to link donors' maternal and paternal pedigrees to known places and ethnicities backwards in time for at least a few centuries. Some day, millions of public DNA sample results with detailed analyses matched to long pedigrees may become accessible, but at the moment, the available DNA test results are strongly skewed towards northern Europe and especially the UK and Iceland (Greenhough 2006; Helgason et al. 2000), and are not linked to in-depth pedigrees.

The promise of future research in human genetics, however, when combined with recent archaeological evidence and a genealogy community anxious to process the latest pertinent data, is to test the premise of ethnicity as a *Blut und Boden* or primordialist proposition. Genetic scholarship to date suggests that site- and group-specific research is needed to determine whether members of a specific ethnic group share a high proportion of genetic markers in common, or whether individuals of diverse pedigrees but shared culture traits obtained the latter principally through diffusion, because different groups may have had very different histories.

Genetic genealogists' interests in finding their roots in a known ethnic group or even clan, where these are unknown through conventional archival work or family tradition, has encouraged DNA testing companies to market imagined communities in the few cases where lineages can be ascertained with reasonable certainty. Today DNA testing companies offer to reveal whether men carry the Pictish gene (Goodwin, 2006), the Viking gene (Renfrew, 2001), Jewish gene (Hammer et al., 2000); or even whether Jewish men's paternal ancestors were Cohens, the biblical temple priests descended from Aaron (Skorecki et al., 1997). Other tests purport to show patilineal descent from Genghis Khan (Zerjal, Xue, et al., 2003; www.familytreedna.com/matchgenghis.html), or Niall of the Nine Hostages of Irish legend (Moore, et. al., 2006; http://www.familytreedna.com/matchnialltest.html). (This Niall was the purported founder of the O'Neill surname and its variants.) These tests have immediate geographical implications for some clients, as the traditional lands of the O'Neill sept can be located on Irish genealogy maps, or inferred from Irish promotional websites like www.GoIreland.com. Similarly, a man with a Cohen gene might reasonably claim descent from biblical Israel.

Results of these commercial tests nevertheless present further complexities for familial or ethnic identities. As in the Irish-Iberian example stated above, they do not guarantee fail-proof ethnicities. The "Jewish gene" on the Y-chromosome also appears among non-Jewish Spaniards and Arabs (Hammer et al., 2000): a common origin for members of all three groups in the Near East, or forced conversions of Spanish Jews may explain these distributions. A "Viking gene" in an Englishman could indicate Scandinavian descent—but not whether his ancestor arrived from Norway during the Dark Ages or from France via the Norman Conquest.

Picts and Vikings, to the extent that their historical identities are even today well understood, vanished from written records during the Middle Ages, leaving only present-day northern Europe and its incompletely preserved heritage sites to be explored by the North American family historian interested in discovering what his paternal-line ancestors' lives were like. Unfortunately for the deep genealogy tourist, Scottish and Norwegian landscapes today are the result principally of the past few hundred years of evolution of the modern nation state and more recent cultural changes, rather than of a deep genealogy-relevant time. The marketing manager of VisitScotland nevertheless hopes that the new tests for the "Pictish gene" will stimulate genealogy tourism to Scotland (Goodwin, 2006; Duffy, 2006).

The Y-DNA debates on popular message boards (for example, of Rootsweb.com at http://listsearches.rootsweb.com/?list=GENEALOGY-DNA) reveal a surprisingly high interest in linking today's spatially clustered Y-DNA sub-groups with ethnic groups chronicled in antiquity, for places where contemporary ethnic distributions are only weakly identifiable. They debate the existence of Friesian, Belgic, Swabian or other ethnically nicknamed markers (haplotype sub-clusters) that are more finely tuned, both genetically and geographically. Their premise is that DNA analysis can reveal ancestors from small ethnic groups who today maintain only minor cultural differences from their larger nation and society. Additionally, they hope that links can be demonstrated to societies absent from European maps today but that were mentioned by classical authors such as Strabo and Ptolemy.

One problem with this line of inquiry, however, concerns the making of untested conclusions from spatial correlations that in and of themselves may only exemplify the fallacy of *cum hoc ergo propter hoc*—that simple correlation indicates causality. For example, genetic markers whose mutation could have occurred thousands of years back into pre-history may not correlate in any causal fashion with ethnic groups that originated more recently; or that were identified in early written records of antiquity but who have vanished from modern maps. Folkloric references to ethnic groups' migration patterns, pedigrees, or sites of origin such as Irish legends of the Milesian kings (Irish Texts Society, 1938) should not be dismissed, but they do require cautious handling as evidence. Incontrovertible links between material culture complexes determined from archaeological sites (such as the La Tène culture of current interest to genetic genealogists in Europe) are notoriously difficult to reconcile with ethnic groups known to historians or modern ethnologists; and certainly to whatever meanings non-literate prehistoric societies may have had about themselves (Megaw and Megaw, 1994).

The reliability of classical geographers like Strabo is also in question. Greek and Roman lists of tribes at the further reaches of their empires were seldom based upon the authors' personal experience; and later authors often plagiarized earlier sources without verifying their information (Jones, 2000; Freeman, 2001; Todd 2004). They did not clearly distinguish Germanic and Celtic cultures, or the role of ancestry vs. shared culture traits in their classifications of specific ethnic groups.

The quest nevertheless continues to link current geographically clustered DNA markers to former tribes, and may yield future success. Twenty years ago, the whole notion of a "Viking gene" was imaginable but indemonstrable.

Imagined Communities of Kinship

More disappointingly for clients who contribute their y-DNA test results to family surname projects in the hopes of demonstrating a clan affinity, perhaps 50% of men tested for y-DNA links to family surnames do not seem to be genetically related to other men with the same surname (Sykes and Irven, 2000). These discrepancies may conceal long-ago "non-paternal events," such as a male ancestor who fathered an illegitimate child, changed his surname, adopted a child from an unrelated family, and so on. A likely cause is simply the small sample size for most surname projects, suggesting that results appearing as random today may later "join" a larger cluster when more records are added. Although many Irishmen have the Naill of the

Nine Hostages genetic marker, few are actually surnamed O'Neill or its variants, suggesting an older "upstream" mutation prior to Niall's birth (McEvoy and Bradley, 2006). Other surnames appear to have multiple origins and discrete DNA clusters. Surnames based upon common personal names (Johnson), occupations (Smith) or topography (Hill) would logically have multiple unrelated paternal founders.

The CASA Surname Profiler (n. d.) data base for Britain, developed by the Centre for Advanced Spatial Analysis of the University College London does suggest that even common surnames exhibit distinctive geographic clusters, especially when the 1881 British census is consulted, rather than recent census data. Yet comparing birth places of individuals listed on the 1881 census with their locales at the time of the census can nevertheless reveal considerable population mobility, notably in the high percentage of Irish-born who emigrated to England in search of employment. (This census is on-line at http://www.familysearch.org/.) Such family migration histories are crucial to the entire project of linking DNA markers with particular places or hypothesized ethnic groups, yet family memories may not retain memories of migration after a century has passed.

Whether due to non-paternal events or exceptionally rapid rates of mutation, the mixed results of surname and Y-DNA projects to date actually subvert, rather than reinforce patriarchal ideologies of the family as a set of individuals related by blood or by sanctioned marriage, as traced through the paternal line. They leave open possibilities of understanding families in ways that are more acceptable to social anthropologists and hybridity theorists (Wade, 2005), who have long construed kinship in terms of social reproduction, rather than the replication of DNA.

Imagined Communities in the Marketplace

Today capitalism is a major force in genetic genealogy's formulae of imagined communities (Simpson, 2000). Its imperative is to generate profits through sales of genealogical products, whether through innovative marketing strategies or through commonplace merchandise that simply reinforces older—and scientifically unsubstantiated—stereotypes. In the past, companies specializing in products for the family historian offered pedigrees; family coats-of-arms (often spurious); and more recently, software for genealogical data storage. Entrepreneurs might foster a sense of community or territory by offering for sale a Scottish clan tartan or customized tour of clients' ancestral villages in Denmark.

One of the most imaginative geographies to emerge from commercial DNA testing was proposed by Cambridge University geneticist and DNA testing company principal Brian Sykes. He noted that mtDNA donors of European descent belonged to one of only seven haplogroups of mitochondrial DNA. Sykes (2001) concluded that Europeans descended from one of seven original female ancestors who lived during the Paleolithic and Neolithic periods, whom he nicknamed the "seven daughters of Eve" and represented as "clan mothers" or "ancestral mothers".

> ... we can not only give you an exact readout of your DNA sequence, but also discover to which of the clans you belong, and from which ancestral mother you are descended. For many of the ancestral mothers, and there are about 36 world-wide, we know whereabouts

they lived and how many ten of thousands of years ago. DNA changes very slowly over time and this is what we use to calculate how long ago the clan mothers lived. By studying features of the geographical distribution of their present-day descendants, we can work out where they lived as well. To emphasise that they were real individuals, we have given them all names and, using archaeological and other evidence, we have reconstructed their imagined lives (http://www.oxfordancestors.com/your-maternal.html).

Sykes's (2001) popular best-seller, *The Seven Daughters of Eve*, links each haplogroup to a fictionalized sketch of each female founder, together with a name derived from the letter symbolizing her haplogroup. H is Helena who lived in the south of France during the Pleistocene, V becomes Velda, who emerged 17,000 years ago in southwestern Spain, and so on (http://www.oxfordancestors.com/your-maternal.html). To further develop and market imagined communities based upon the *Seven Daughters*, Oxford Ancestors offers on-line shopping for the book, a MatriMap™, and a color certificate showing the genetic affinities of the seven haplogroups with their multiple un-named descendant sub-groups, with a star on the purchaser's affiliate sub-group. Also for sale is a MotherStone™ pendant of particular variety of semi-precious stone jewelry that is pegged to each of the ancestral mtDNA haplogroups. In effect, Oxford Ancestors constructs new imaginative geographies and communities to replace the old ethnicity based upon language, food customs, and textbook map boundaries. New maps and pendants now link the customer to a fantasized foremother in a Pleistocene or Neolithic homeland with little relationship to modern nationalities.

The IBM web pages on their partnership in the National Geographic Society Genographic Project attempt to link prospective DNA donors to an imagined community of scholars. Initially IBM encourages the reader to view the entire scope of human migration and settlement in personal terms: a kind of "brotherhood of Man" trope (http://www-03.ibm.com/industries/healthcare/pdfs/genoflash/data/html/index.html). IBM also solicits DNA donors by inviting them to construe themselves as part of the actual research group: "By becoming involved in the Genographic Project through the purchase and use of the participation kit, you are in a sense becoming an "associate explorer' on our team—a rare and exciting opportunity to get involved in a major scientific endeavor..." (http://www-03.ibm.com/industries/healthcare/genographic/doc/content/landing/144492313...)

The other principal English-language DNA testing firms' web sites reveal less imaginatively promoted DNA tests, reports, contact information, and surname projects. The larger market place for genealogy products, however, is now augmenting DNA testing through a predictable array of gift items (see the on-line Genealogy Gifts Store, viewed at http://www.cafepress.com/jmkbooks.) The question, however, is whether a commercialized genetic genealogy will have consequences in the political arena as warned by Simpson (2001) and Nash (2004, 2006), beyond clients' personal interests in their DNA test results, notably if its branding strategies depend upon marketing gimmicks.

Conclusions

Because the practices and findings of genetic genealogy are so rapidly evolving, any definitive conclusions today are likely to become outmoded tomorrow. But precisely the point of this paper is that deep genealogy does not easily lend itself to single authoritative readings. The different backgrounds, motives, and experiences of deep genealogists enable multiple and contradictory constructions of meaning of raw DNA test results. The linking of one's DNA to modern-day surname projects in Ulster or to an indigenous Gaelic-speaking population in prehistory will likely appear very differently to a 10th-generation resident of Belfast than they will to an assimilated American with forgotten Ulster roots. Moreover, genetic genealogy better meets the definition of "communities of practice" rather than of material communities located in particular places and cultures (Wenger, 2001). Unlike older, hegemonic narratives of nationality and ancestry, Internet web sites and message boards convey a passionate, ongoing debate about the meanings of genetic markers, ethnicity, homeland, belonging, kinship and identity; in which the genealogical past is indeed contested space.

References

Anderson, B. (1991), *Imagined Communities*, Verso, London.
Brodwin, P. (2002), "Genetics, Identity, and the Anthropology of Essentialism", *Anthropological Quarterly*, 75, pp. 323-30.
CASA Surname Profiler, Center for Advanced Spatial Analysis, University College, London <http://www.spatial-literacy.org/UCLnames/Surnames.aspx>
Cavalli-Sforza, L. L., Menozzi, P. and Piazza, A. (1994), *The History and Geography of Human Genes*, Princeton University Press, Princeton.
Cohen, R. (1997,) *Global Diasporas*, Routledge, London.
Coles, T. and Timothy, D. J.(eds.) (2004), *Tourism, Diasporas, and Space*, Routledge, London.
Dodge, M. and Kitchin, R. (2001), *Mapping Cyberspace*, Routledge, London.
Duffy, J. (2006), "New Genealogy Centre to Create DNA Library", *Sunday Herald* 29 October, (Scotland), < www.sundayherald.com>.
Edensor, T. (2002), *National Identity, Popular Culture and Everyday Life*, Oxford, New York.
Fleising, U. (2001), "Genetic Essentialism, *Mana*, and the Meaning of DNA", *New Genetics and Society*, 20, pp. 43-57.
Freeman, P. (2001), *Ireland and the Classical World*, University of Texas Press, Austin.
Friedman, J. (1994), *Cultural Identity and Global Processes*, Sage, London.
Goodwin, K. (2006). "Think you're a real Scot? Try checking your DNA", *The Sunday Times*, 13 August <www.timesoneline.co.uk>
Graham, B.J, Ashworth, G.J. and Tunbridge, J.E. (2000), *A Geography of Heritage: Power, Culture and Economy*, Arnold, London.

Gramsci, A. (1971), *Selections from the Prison Notebooks*, Lawrence and Wishart, London.
Greenhough, B. (2006), "Tales of an Island-Laboratory: Defining the Field in Geography and Science Studies", *Transactions of the Institute of British Geographers*, 31, pp. 224-237.
Greenhough, B. and Roe, E. (2006), "Towards a Geography of Bodily Technologies", *Environment and Planning A*, 38, pp. 416-422.
Hammer, M.F., Redd, A.J., Wood, E.T., Bonner, M.R., Jarjanazi, H., Karafet, T., Santachiara-Benerecetti, S., Oppenheim, A., Jobling, M.A., Jenkins, T., Ostrer, H. and Bonne-Tamir, B. (2000), "Jewish and Non-Jewish Middle Eastern Populations Share a Common Pool of Y-Chromosome Biallelic Haplotypes", *Proceedings of the National Academy of Science*, 97, pp. 6767-6774.
Helgason, A., Hrafnkelsson, B., Gulcher, J.R., Ward, R., and Stefánsson, K. (2003), "A Populationwide Coalescent Analysis of Icelandic Matrilineal and Patrilineal Genealogies: Evidence of a Faster Evolutionary Rate of mtDNA Lineages than Y Chromosomes", *American Journal of Human Genetics*, 72, pp. 1370-1388.
Helgason, A., Sigurðardóttir, S., Gulcher, J., Stefánsson, K. and Ward, R. (2000), "Sampling Saturation and the European mtDNA Pool: Implications for Detecting Genetic Relationships among Populations", in C. Renfrew and K. Boyle (eds), *Archaeogenetics: DNA and the Population Prehistory of Europe*, McDonald Institute for Archaeological Research, Cambridge, 285-294.
Heyer E., Puymirat J., Dieltjes P., Bakker E., de Knijff, P. (1997), "Estimating Y chromosome specific microsatellite mutation frequencies using deep rooting pedigrees", *Human Molecular Genetics*, 6(5): 799-803.
Hill, E.W., Jobling, M.A., Bradley, D.G. (2000a), "Y-chromosome variation and Irish origins: A Pre-Neolithic Gene Gradient Starts in the Near East and Culminates in Western Ireland", *Nature*, 204: 351-252.
Hill, E.W., Jobling, M.A. and Bradley, D.G. (2000b), "Y-chromosome variation and Irish origins", in C. Renfrew and K. Boyle (eds), *Archaeogenetics: DNA and the Population Prehistory of Europe*, McDonald Institute for Archaeological Research, Cambridge, pp. 203-208.
Ignatieff, M. (1994), *Blood and Belonging: Journeys into the New Nationalism*, Farrar, Straus, and Giroux, New York.
Irish Texts Society (1938), *Lebor Gabala Erenn (The Book of the Taking of Ireland)*, Educational Company of Ireland, Dublin.
Jobling, M.A. (2001), "In the Name of the Father: Surnames and Genetics", *Trends in Genetics*, 16, pp. 353-357.
Jones, A. (2000), *Ptolemy's Geography*, Princeton University Press, Princeton.
Leeds-Hurwitz, W. (ed.) (2005), *From Generation to Generation: Maintaining Cultural Identity over Time*, Hampton Press, Inc., Cresskill, NJ.
McCarthy Surname Study (n.d.) <http://www.familytreedna.com/(wb13yg3etdikuz55s5mbvh2s)/public/McCarthySurnameStudy/index.aspx> viewed July, 2006.
McEvoy, B. and Bradley, D.G. (2006), "Y-chromosomes and the Extent of Patrilineal Ancestry in Irish Surnames", *Human Genetics*, 119, pp. 212-219.

McEvoy, B. and Edwards, C.J. (2005), "Reappraising the Viking Image", *Heredity,* 95, pp. 111-12.
Megaw, R. and Megaw, V. (1994), "Through a Window on the European Iron Age Darkly: Fifty Years of Reading Early Celtic Art", *World Archaeology,* 25, pp. 287-303.
Meethan, K. (2004), "'To stand in the shoes of my ancestors': Tourism and Genealogy", in T. Coles and D. J. Timothy (eds), *Tourism, Diasporas, and Space,* Routledge, London, pp. 139-150.
Moore, L.T., McEvoy, B., Cape, E., Simms, K., Bradley, D.G. (2006), "A Y-Chromosome Signature of Hegemony in Gaelic Ireland", *American Journal of Human Genetics,* 78, pp. 334-338.
Nash, C. (2002), "Genealogical Identities", *Environment and Planning D: Society and Space,* 20, pp. 27-52.
Nash, C. (2004), "Genetic Kinship", *Cultural Studies,* 18, pp. 1-33.
Nash, C. (2005), "Geographies of Relatedness", *Transactions, Institute of British Geographers,* n.s. 30, 449-462.
Nash, C. (2006), "Irish Origins, Celtic Origins: Population Genetics, Cultural Politics", *Irish Studies Review,* 14, pp. 11-37.
Nelkin, D. and Lindee M.S. (2004), *The DNA Mystique: the Gene as a Cultural Icon,* University of Michigan Press, Ann Arbor.
Oppenheimer, S. (2006), *The Origins of the British: A Genetic Detective Story: The Surprising Roots of the English, Irish, Scottish, and Welsh,* Carroll and Graf Publishers, New York.
Palsson, G. (2002), "The Life of Family Trees and the Book of Icelanders", *Medical Anthropology,* 21, pp. 337-67.
Parsons T. J., Muniec, D. S., Sullivan, K., Woodyatt, N., Alliston-Greiner, R., Wilson, M. R., Berry, D. L., Holland, K. A., Weedn, V. W., Gill, P. and Holland, M. M. (1997), "A High Observed Substitution Rate in the Human Mitochondrial DNA Control Region," *Nature Genetics,* 15: pp. 363-367.
Pomery, C. (2004), *DNA and Family History,* The Dundurn Group, Toronto.
Rabinow, P. (1999), *French DNA: Trouble in Purgatory,* University of Chicago Press, Chicago.
Renfrew, C. (2001), "From Molecular Genetics to Archaeogenetics", *Proceedings of the National Academy of Science,* 98, pp. 4830-4832.
Renfrew, C. and Boyle, K. (eds.) (2000) *Archaeogenetics: DNA and the Population Prehistory of Europe,* McDonald Institute for Archaeological Research, University of Cambridge, Cambridge.
Richards, M., Corte-Real, H., Forster, P., Macaulay, V., Wilkinson-Herbots, H., Demaine, A., Papiha, S., Hedges, R., Bandelt, H.J. and Sykes, B. (1996), "Paleolithic and Neolithic Lineages in the European Mitochondrial Gene Pool", *American Journal of Human Genetics,* 59, pp. 185-203.
Romanucci-Ross, L., De Vos, G.A. and Tsuda, T. (eds.) (2006), *Ethnic Identity: Problems and Prospects for the Twenty-first Century,* 4th ed., Altamira Press, Lanham.

Rootsi, S., Magri, C., Kivisild, T., Benuzzi, G., Help, et al. (2004), "Phlygeography of Y-Chromosome Haplogroup I Reveals Distinct Domains of Prehistoric Gene Flow in Europe", *American Journal of Human Genetics* 75(1): pp. 128-137.

Rosenberg, N.A., Pritchard, J.K. et al. (2002), "Genetic Structure of Human Populations", *Science*, 298, pp. 2381-2385.

Santos, R.V. and Maio, M.C. (2004), "Race, Genomics, Identities and Politics in Contemporary Brazil", *Critique of Anthropology*, 24, pp. 347-378.

Simpson, B. (2001), "Imagined Genetic Communities: Ethnicity and Essentialism in the Twenty-first Century", *Anthropology Today*, 16, pp. 3-6.

Shriver, M.D. and Kittles, R.A. (2004), "Genetic Ancestry and the Search for Personalized Genetic Histories", *Nature Reviews Genetics*, 5, pp. 611-618.

Skinner, D. (2006), "Racialized Futures: Biology and the Changing Politics of Identity", *Social Studies of Science*, 36, pp. 459-488.

Skorecki, K., Selig, S., Blazer, S., Bradman, R., Bradman, N., Waburton, P.J., Ismajlowicz, M. and Hammer, M.F. (1997), "Y Chromosomes of Jewish Priests", *Nature*, 385 (2), p. 32.

Smolenyak, M.S. and Turner, A. (2004), *Trace Your Roots with DNA: Using Genetic Tests to Explore Your Family Tree*, Emmaus, PA, Rodale.

Strathern, M. (1995), "Nostalgia and the New Genetics", in D. Battaglia (ed.), *Rhetorics of Self-making*, University of California Press, Berkeley, pp. 97-120.

Sykes, B. (2001), *The Seven Daughters of Eve*, Norton, New York.

Sykes, B. (2006), *Saxons, Vikings and Celts: the Genetic Roots of Britain and Ireland*. New York: W.W. Norton Co., Inc.

Sykes, B. and Irven, C. (2000), "Surnames and the Y Chromosome", *American Journal of Human Genetics*, 66, pp. 1417-1419.

Taylor, G. and Spencer, S. (eds.) (2004), "Introduction", in *Social Identities: Multidisciplinary Approaches*, Routledge, London.

Timothy, D.J. and Teye, V.B (2004), "American Children of the African Diaspora", in T. Coles and D. J. Timothy (eds), *Tourism, Diasporas, and Space*, Routledge, London, pp. 111-123.

Todd, M. (2004), *The Early Germans*, 2nd edition, Blackwell, Oxford.

Tunbridge, J.E. and Ashworth, G.J. (1996), *Dissonant Heritage: The Management of the Past as a Resource in Conflict*, J. Wiley, New York.

Tutton, R. (2004), ""They Want to Know Where They Came From": Population Genetics, Identity, and Family Genealogy", *New Genetics and Society*, 23, pp. 106-120

Underhill, P. A., Shen, P., et al. (2000), "Y Chromosome Sequence Variation and the History of Human Populations", *Nature Genetics*, 26, pp. 358-61.

Vallone, P.M. (2003), Development of Multiplexed Assays for Evaluating SNP and STR Forensic Markers. http://www.cstl.nist.gov/div831/strbase/pub_pres/ValloneGWU2003.pdf

Wade P, (2005), "Hybridity Theory and Kinship Thinking", *Cultural Studies*, 19, pp. 602-621.

Weale, M.E., Weiss, D.A. et al. (2002), "Y Chromosome Evidence for Anglo-Saxon Mass Migration", *Molecular Biology and Evolution*, 19, pp. 1008-1021.

Weinreich, R., Bacova, V. and Rougier, N. (2003), "Basic Primordialism in Ethnic and National Identity", in P. Weinreich and W. Saunderson (eds), *Analysing Identity: Cross-cultural, Societal and Clinical Contexts*, Routledge, London, pp. 115-169.

Wells, S. (2006), *Deep Ancestry: Inside the Genographic Project*, Washington, D.C.: National Geographic Society.

Wenger, E. (2001), "Communities of Practice", in N.J. Smelser and P.B. Baltes (eds), *International Encyclopedia of the Social and Behavioral Sciences*, pp. 2339-42.

Williams, S.R. (2005), "A Case Study of Ethical Issues in Genetics Research: the Sally-Hemings-Thomas Jefferson Story", in T.R. Turner (ed.), *Biological Anthropology and Ethics: from Repatriation to Genetic Identity*, State University of New York Press, Albany, pp. 185-208.

Wilson, J.F., Weiss, D.A., et al. (2001), "Genetic Evidence for Different Male and Female Roles During Cultural Transitions in the British Isles", *Proceedings of the National Academy of Science*, 98, pp. 5078-5083.

Winston, C.E. and Kittles, R.A. (2005), "Psychological and Ethical Issues Related to Identity and Inferring Ancestors of African Americans", in T.R. Turner (ed.), *Biological Anthropology and Ethics: from Repatriation to Genetic Identity*, State University of New York Press, Albany, pp. 209-229.

Zerjal, T., Xue, Y., et al. (2003), "The Genetic Legacy of the Mongols", *American Journal of Human Genetics* 72, pp. 717-21.

Chapter 10
Conclusion: Personal Perspectives

Dallen J. Timothy and Jeanne Kay Guelke

Perhaps the most conspicuous thread throughout all nine preceding chapters is the personal identity quest involved in doing genealogy and family history. Place is also essential to family history research. People lived, worked, and played in places. Throughout history they migrated from place to place, building trails of homelands and multiple place-bound identities. While the primary focus of genealogy and family history is people, the importance of ancestral connections to place is unmistakable.

The spatial and temporal tools used to unravel histories, document personalities, and chronicle events and places are valuable to family historians. Similarly, family history is a rich topic of research that ought to be of significant interest as a field ready to harvest by the toolset that geography provides. The desire to connect with peoples and places of the past motivates people to discover their roots directly in the homelands of their forebears as "roots tourists" (Kivisto, 2003; Stephenson, 2002; Timothy, this volume), through virtual spaces and cyber ethnic communities on the Internet (Chan, 2005; Crowe, 2003; Meethan, this volume), or via old letters, archives, photographs, and living family gatekeepers (Blaikie, 2001; Gait, 2006; Richards, this volume). Likewise, maps are crucial in documenting homelands, locating ancestral properties, discovering churches and other archival data sources, and understanding regional/national boundary changes and therefore alternating citizenships and national identities (e.g. Alsace and Prussia) (Kashuba, this volume; Lamble, 2000; Plante, 2003). Some turn to DNA testing to supplement documentary sources about their ethno-spatial origins (Mountain and Guelke, this volume; Nelkin and Lindee, 2004; Ruitberg et al., 2001). Likewise, governments and genealogical associations are beginning to realize the need to protect potential genealogical resources via cultural resource management and environmental impact assessments (Hunter, this volume; Smith, 2000; Tull, 2004). And for many people, regardless of their methods, genealogy brings spiritual rewards or fulfills religious obligations (Davies, 2000; Ludlow, 1992; Otterstrom, this volume).

These issues are not only salient for scholarly researchers but for genealogy practitioners as well. As editors of this volume, we believe that researchers must look beneath the surfaces of research *about* family history and realize that each family historian *comes from* a particular past and approaches her or his topic in deeply personal ways. As Kevin Meethan suggests, genealogy has a profoundly idiosyncratic and perhaps even lonely dimension, despite its basis in the extended family, because only siblings share an identical set of roots. Yet even siblings do not share the same experiences of self in relation to family and may not have similar passions for doing family history. Birth order affects a child's personality development, and in the past (and present in some parts of the world), it determined his or her inheritance. Gender goes without saying. Children may be named after grandparents as a means of sustaining an ancestor's personal legacy, or be named after nobody at all as a means of breaking the links to the past. Family history is thus simultaneously both a collective and uniquely personal endeavor. The previous chapters emphasized general trends and principles. In this afterword, however, we extend these discussions to incorporate the personal.

To conclude this volume we decided to take a more personal approach to family history by way of giving the uninitiated reader a better sense of the meaning of family history to its practitioners. For example, behind Sam Otterstrom's review of Latter-day Saint genealogy is a set of deeply held religious convictions. Penny Richards' chapter builds upon her own family's collection of nineteenth and early twentieth-century trans-Atlantic correspondence. Bill Hunter's memories of his grandfather are only superficially distanced from his current work as a consultant, because they allow him to make important connections between family, local history, and culturally sensitive landscapes.

Human geographers' current concerns with reflexivity and situated knowledge confirm that there is no "view from nowhere." We consequently asked our authors to provide a brief statement about how their personal interests in family history influenced their academic research; or alternatively, how their professional expertise informed their family history pursuits. Nevertheless, the following autobiographical vignettes offer the reader an immediate sense of the meaning of family history in individual lives, which underlay the writings in the preceding chapters.

Bill Hunter

As a little boy, I loved to play cops and robbers, and would dress the part, complete with a hat, holster and badge. It was the badge that set my uniform apart, because this shiny metal star impressed with "Deputy Sheriff" was real, a gift from my grandfather, then in his twilight. A kind and gentle man, his smile and the twinkle in this eye was, to me, the embodiment of an "Irish" grandfather, the underdog who, through hard work and grit, had emerged from the steel mills and coal mines to lead his family into the American middle class. I fashioned my identity around my warm memory of this good man, and around the personal histories and memories of my Mom and her brothers.

Like many children of the American suburbs in the late 1960s and early 1970s, the mythology of my ancestry played an important part in my development of a personal narrative. Growing up in the polyglot suburbs of Cleveland, Ohio, my family's self-fashioned Irishness, performed within the family or through the annual St. Patrick's Day celebrations, set me apart from the Armenian, German, Polish, Russian and "white American" identities of my playmates. And yet, just as many of my friends displayed no outward ethnic identity and had never created an ancestral mythology, the long struggles of their family's travels to the comfort of the eastern suburbs demanded a de-emphasis of their particular genealogies.

As so often happens, as we grow and seek our own identity, we find that the mythologies of our ancestry clash with the complexities of our genealogies. My encounter with the mythologies began in earnest with my work as a young field historian, conducting cultural resource assessment surveys of landscapes and places tied closely to both my ancestry and genealogy. Tasked to survey a highway corridor between Cadiz and St. Clairsville, Ohio, in the eastern Ohio coal fields, my encounter with the landscape and the histories embedded within it challenged my sense of self.

As my fieldwork and research progressed, I had the opportunity to begin to visit the places vaguely outlined in the family narrative, and slowly, the partial images and memories that I have considered to be fanciful began to emerge from the landscape and literature. The stories of the armed guards, roadblocks, mobs of angry coal miners, and violence rushed to join the image of the politically progressive, faithfully Catholic, and thoroughly Democratic Irish-American embodied by my grandfather.

In the spring of 1931, with the radical National Miners Union organizing among the Hungarian and Slovak miners of the eastern Ohio coalfields during the dark early days of the Depression, the Irish foreman, inspectors and merchants—acculturated through the Church, fraternal organizations and veterans groups—knew what side they were on. Armed by the coal company and deputized by the county, my grandfather helped in the vigorous suppression of the radical union and the breaking of a particularly bitter strike. The one genealogical line and family narrative elevated above all others had eliminated a tremendously important historical moment from its memory.

This stunning realization was the first crack in the mythology of my ancestry, and I came quickly to appreciate the duplicity of landscape and the problems of personal mythology. As my later work led me to other places and into other landscapes once occupied by ancestors from the less mythologized or even unmentioned branches of the family tree, I encountered the more troubling genealogies: My father's father dying of syphilis in the notorious Newburgh State Hospital in 1923 rather than being hit by a streetcar as my father believed; the lionized great-grandfather not in fact being directly related at all, but rather acting as a stepfather to a girl orphaned by a mine collapse in Arkansas and sent to "her people" in Ohio by her Mississippi-born mother—a genealogical line falling well outside the narrative of the northern industrial heritage. The dead ends, absences and silences, the suicides and untimely deaths, the hard choices and hidden tragedies help to explain my enthusiastic embrace of a single, straightforward Irish identity.

Somewhere, buried in the bushes or in the cul-de-sac of my childhood street, is that badge, a scrap of history that rent my ancestral mythologies to expose a no less proud but infinitely more complex, and immensely enriching, family history.

Melinda Kashuba

I became interested in genealogy when I moved to UCLA to pursue my doctorate in geography. An elderly cousin took me under her wing and generously shared the information she had inherited from her mother and our mutual grandmother about our family. As it turned out, a fellow graduate student at UCLA was an avid genealogist. She and I shared discoveries made on our respective lines in furtive conversations. We shared this little secret of ours for several years. She admonished me *never* to let faculty members know that I was pursing genealogy because I would not be taken seriously as a researcher. Twenty-five years ago, genealogy was poorly regarded as a serious subject for study by university faculty. I think it had the image of a vocation associated with "little old ladies in tennis shoes" rather than a serious subject to be taken up as research in an academic setting. Curiously, academic historians shared this view also until they discovered the trove of information available in oral histories, family papers, and self-published family histories that lent texture to their academic work.

I finished my degree and while teaching geography at a local college became deeply involved with local and regional genealogy societies. In the spirit of wanting to give back to the genealogical community that had assisted me in my own personal genealogical work, I gave lectures on migration, maps, and library research strategies during that decade after leaving UCLA to local, regional and national societies. In every instance I found an audience of genealogists thrilled with this "new" information that might assist them in locating their ancestors. Would that my college geography students have been so attentive and interested in what I had to say!

Three years ago, I was accepted as a participant in a National Endowment for the Humanities sponsored Summer Institute on "Reading Popular Cartography" held at the Newberry Library in Chicago. Although I had worked almost exclusively in the arena of genealogy for the previous ten years as an independent researcher, writer and lecturer, I still felt that I was a geographer who happened to be researching, writing and lecturing about genealogical subjects. That Institute rekindled my love for geography and in conversations with other Institute participants and guest faculty, I was able to communicate to them that there are a great number of "lay persons" eager to hear what geographers had to say about everything from how to locate maps to reading the cultural landscape in a way that might provide them clues to their ancestral roots.

Since that Institute, I wrote a book about using American maps in genealogical research. The book has been generally well-received by genealogists and has encouraged me to continue writing about geography for the genealogical community. Recently, I began teaching college geography again and one of my goals was to connect students' lives in a meaningful way to the concepts taught to them through lecture and assigned reading. In my cultural geography class I have assigned students a "Routes and Roots" term paper that requires them to trace their family history along one family line in an effort to apply geographic concepts related to migration (legal and illegal), settlement patterns, "cultural baggage", and economic development to their personal lives. They are required to submit a pedigree chart along with maps at various scales showing places of vital events and possible migration routes of their family. Past students have told me that this assignment caused them initially great anxiety but ultimately became one of their favorite college assignments because it challenged them to apply some of what they had learned in class to their personal situations.

Jeanne Kay Guelke

As an historical geographer, I found the adoption of family history as a hobby to be both natural and startling. I am used to thinking passionately about people and landscapes of the past, an interest spurred by some elderly relatives who were happy to talk about major historical events that they had experienced, and who lived in places like rural New England where "history on the ground" was visible even to children. I knew only vaguely about ancestors who "came from Germany" or "from Ireland and Quebec," and a few relatives gave me some short pedigrees that I tucked away in a drawer and seldom examined. Surprisingly, I never thought seriously to learn about my ancestors until I reached late middle age. My mother's extended family were planning a reunion, and I thought they might be interested in seeing some large-scale maps of ancestral villages, even though I previously had never bothered to identify them and knew nothing about their landscapes. Once I found the topographic maps, I was shocked to realize that my ancestors' "German" villages along the North Sea were extensively diked and drained. I took another look at one of the pedigrees, and even my bad high school German was sufficient to determine that a major group of these ancestors did not have German or Dutch names. Nobody had ever hinted that my late grandmother's people were Friesians—a group I knew nothing about and whose culture was unknown to me.

I began to realize that I had a genetic and cultural past of which I was ignorant, and that much of what I considered to be my unique identity or contemporary social categories in fact derived from my ancestors' DNA (blue eyes, light skin) and from the many choices they made long before I was born, such as dropping their religious affiliations, emigrating to the United States, and deliberately assimilating into its mainstream culture. Want it or not, I had a "genealogical identity" (Nash, 2002).

My brother, David Mountain, became an avid family historian at about the same time, and is now a "serious amateur" with an on-line surname project and subscriptions to several genealogical data bases. When his search for paternal ancestors "hit the brick wall" he pursued DNA testing, learning the genetics necessary to analyze both his test results and those of prospective distant relatives. Out of our shared interest in family history, our sibling relationship has changed from one of distant cordiality to frequent e-mail correspondence about recent questions and findings.

Kevin Meethan

My involvement in genealogy began as the result of a phone conversation with my mother. "I'm going to research our family tree" she said. "Good luck" I replied, and thought little of it until a week or so later, when she phoned again. "You've got a computer and Internet access, can you look up some stuff for me?" And that was it. In part I viewed it as a challenge to my research skills, and partly I was also fascinated by the possibilities that the Internet offered. There were two other factors at play. First, as a small child I had been intrigued by a collection of old sepia photographs that an aunt of mine had, and which, since her death, had been stored in my attic. Out they came, and my mother also gave me some she had been keeping, which I was later able to date back to the mid-nineteenth century. Second was the origin of my surname, something that had also intrigued my father. So while I set about trying to connect faces to names, places and times, trawling the archives and browsing message boards for connections, trying to work around the variations of my surname, I came across emails in which people talked about travelling back to the land of their ancestors. On the academic side, as someone who has an interest in tourism and cultural change, and a background in social anthropology from which I knew the importance of kinship in relation to identity, the connection between my personal interests and academic ones was simple enough to make.

Samuel Otterstrom

As a member of the Church of Jesus Christ of Latter-day Saints, I have a natural interest in knowing the historical context of family history research within the Church. Genealogy research has been a longstanding tradition in my own family partly because of my nineteenth-century Mormon pioneer ancestry. Over time, records of my forebears gathered by relatives and others have come into my interested hands. Years ago I remember being very impressed with a pedigree chart that my grandfather compiled that ostensibly followed one chain of ancestors all the way to Adam in the Old Testament. In turn, Mormon temple work is also familiar to me. When I was a child my parents often made long trips from Spokane, Washington, to the nearest temple in Cardston, Alberta, while my first temple excursion was to the Idaho Falls, Idaho, Temple when I was twelve. Since that time I have had many experiences with both doing genealogical research and in attending LDS temples across the country. Writing this chapter has enlightened me concerning the development of genealogical and family history research among Mormons and it has afforded me the opportunity and challenge to write for an audience unfamiliar with LDS beliefs and practices. On a related note, in my academic pursuits in historical population and settlement geography I have discovered some of the potential uses of genealogical data in reconstructing past migrations and spatial relationships within families, which makes this volume of double interest.

Penny L. Richards

In 1994, I sent to several branches of my family a transcription of some 250 letters sent to our great-great grandmother. (The Marion Brown letters I describe in my chapter are among those.) In 2001, I gave the same branches a CD of about 200 photos in that same great-great grandmother's albums. Then, just a few weeks ago (March 2007), I got a call from a long-lost cousin—we last met more than thirty years ago, when we were little children. Her own son has a school project—the usual family tree, immigration history assignment. She called a cousin, who knew from another cousin about my work with the family papers. It didn't take long for her to track me down online, and soon we were calling and emailing, sending pictures and questions back and forth.

So just now I'm very appreciative of the power of family history. Preserving and studying these documents has been mostly an academic experience for me, but I'm never far from knowing that it's a personal experience too, with personal consequences.

I also had a chance to talk about Marion Brown to a support group for people with MS—Marion had a complicated chronic illness that had a lot in common with multiple sclerosis. I read them excerpts from her letters, showed them photos, explained her predicament as a left-behind cousin in the great wave of emigration. They immediately embraced her story, they laughed, they nodded in recognition, and they asked the kind of questions you rarely hear at an academic conference: did she go to church? Did she have a boyfriend? These are really good questions! And their enthusiasm reminded me that her story cannot just live in the pages of journals and books—because these are stories that people *need*, and they're stories that can only be told through the documents of family history.

Mary Ruvane

While working towards a master's degree in Information Science, at The University of North Carolina at Chapel Hill, I became fascinated with Geographic Information Systems, especially their adoption by disciplines outside of physical geography. My first introduction to GIS for historical research was through an archeology class that focused on technology tools (e.g., GRASS, ArcGIS, ERDAS, digital data, aerial photography, GPS, etc.), culminating in a project that involved mapping visible remnants of the former Indian Trading Path as it traversed sections of present day Orange County, North Carolina. During this time I began to study my own family's history and pondered how best to research and map the historic locations they had settled. What interested me was how few library and archive institutions were aware of the potential GIS held for improving access to their collections and at the same time for enhancing the value of their existing holdings. In my mind, genealogists were the key to making this happen.

Fortunately I was able to merge my two interests after being introduced to G. Rebecca Dobbs, at the time a PhD candidate in the Geography Department at The University of North Carolina at Chapel Hill. She was studying the same Indian Trading Path to understand its influence on shaping present-day town locations in North Carolina's Piedmont region. She needed a method for collecting evidence from 18th century land records to facilitate mapping the neighborhoods that evolved over the course of her study period. We agreed to collaborate, and over the next few years, in exchange for building her a database to collect evidence, I was able to observe the process of transforming historical data into GIS representations of prior neighborhoods and to study their utility for genealogical research and improving access to archive collections.

Dallen J. Timothy

Growing up in the 1970s and 1980s I was intrigued by the fact that my grandfather, Joshua Timothy, someone so close to me in inter-generational terms, was born in the 1800s. The grandparents of everyone else in my social realm were born after the turn of the century. I was very proud of my unique, nineteenth-century Grandpa. When he died in 1988 at age 92, I began to realize how little I really knew about him or his parents. I was especially intrigued by his father, Alma Nephi Timothy, who emigrated from Wales to the United States as a young boy with his parents, but I knew very little about them. My maternal grandmother, Fern Ruesch DeMille, talked a great deal about her father, Ulrich Ruesch, who migrated from Switzerland to the U.S. as a young man in the late-1800s. I relished the few and far-too infrequent stories she shared about his life. Grandma Fern passed away in 1995. In the intervening years since my grandparents' passing, my regret for not delving deeper into their lives while they were living, and my simultaneous desire for knowledge about them and their forebears have intensified. Fortunately, both of these grandparents and their spouses left life histories for their grandchildren to read.

Thanks to online sources and relatives who also value family history, my knowledge has increased. I have since learned much about my ancestors, and in the fall of 2006, copies of my paternal grandparents' courtship letters were made available. What a treasure! One of my most exciting recent discoveries is a website that includes the life history of my great-great grandfather and photographs of the homes where the Timothys lived and churches where they were married and worshipped in Wales. I also recently learned about various European roots on my mother's side that I never knew existed—Norwegian, Dutch, and French. Through one line, the generations have been ostensibly traced through Old King Cole and Charlemagne as far back as 6 AD. Unfortunately, our family has reached a dead end only four generations back from Grandpa Timothy.

I have been blessed with several opportunities to visit Switzerland and Wales, the two "homelands" in my lineage that most intrigue me. Additionally, in recent months I have had opportunities to visit and photograph the graves of several ancestors in Utah and Arizona. Since the 1995 death of Grandma Fern, I have felt a sense of urgency in doing family history and protecting knowledge and remnants of the past that can be preserved. Each new discovery and each site visit has touched me in different ways. Often these are spiritual experiences, but most of all they are marked by a feeling of completeness and awareness that I belong to something greater than myself.

Final Thoughts

As this book demonstrates, genealogy is highly geographical. Social and cultural geographers should have an interest in the social structures of migrant families and the kinship bonds that were often created over long distances. Economic geographers may take interest in economics-based migrations and the capital transfers throughout the world as family history becomes more accessible and globalized, and as diasporic people often sponsor relatives across the world through remittances. Behavioral and tourism geographers have a clear interest in how people travel physically through space and time to their homelands and the experiences other people undergo as virtual globetrotters online. Genealogy also provides a reservoir of possibilities for political geographers and cartographers, as already noted, when political boundaries change, successive nationalisms evolve, and patriotic loyalties adjust. Finally, family history is a natural fit with historical geography with its matching approaches, paradigms, and tools.

Despite the obvious connections between geography and genealogy, genealogy should not belong only to the domain of geography. All of the social sciences (and many physical sciences) have an important role to play in this fast growing trend throughout the western world. Genealogy is now a salient socio-cultural, economic, and political force in the world today and should be considered more seriously as an area of scholarship for psychologists, political scientists, anthropologists, historians, economists, and sociologists. While we hope that it becomes somewhat institutionalized in this regard, our hope remains that it continues to be a passion for millions of people throughout the world who seek to discover their own personal pasts.

References

Blaikie, A. (2001), "Photographs in the Cultural Account: Contested Narratives and Collective Memory in the Scottish Islands", *Sociological Review*, 49(3): 345-367.

Chan, B. (2005), "Imagining the Homeland: The Internet and Diasporic Discourse of Nationalism", *Journal of Communication Inquiry*, 29(4): 336-368.

Crowe, E. P. (2003), *Genealogy Online*, 7th Edition, Emeryville, CA: McGraw-Hill/Osborne.

Davies, D. J. (2000), *The Mormon Culture of Salvation*, Aldershot, UK: Ashgate.

Gait, M. (2006), *Family History Resources in Special Libraries and Archives*, Aberdeen, UK: King's College.

Kivisto, P. (2003), "Social Spaces, Transnational Immigrant Communities, and the Politics of Incorporation", *Ethnicities*, 3(1): 5-28.

Lamble, W.H. (2000), "Genealogical Geography: Place Identification in the Map Library", *INSPEL*, 34(1): 40-51.

Ludlow, D.H. (1992), *Encyclopedia of Mormonism*, New York: Macmillan.

Nash, C. (2002), "Genealogical Identities", *Environment and Planning D: Society and Space*, 20: 27-52.

Nelkin, D. and Lindee, M. S. (2004), *The DNA Mystique: The Gene as a Cultural Icon*, Ann Arbor: University of Michigan Press.

Plante, T.K. (2003), "Enhancing Your Family Tree with Civil War Maps", *Prologue*, 35(2): 53-61.

Ruitberg, C. M., Reeder, D. J. and Butler, J. M. (2001), "STRBase: A Short Tandem Repeat DNA Database for the Human Identity Testing Community", *Nucleic Acids Research*, 29(1): 320-322.

Smith, L. (2000), "'Doing Archaeology?': Cultural Heritage Management and Its Role in Identifying the Link between Archaeological Practice and Theory", *International Journal of Heritage Studies*, 6(4): 309-316.

Stephenson, M. L. (2002), "Travelling to the Ancestral Homelands: The Aspirations and Experiences of a UK Caribbean Community", *Current Issues in Tourism*, 5(5): 378-425.

Tull, S. W. (2004), "Conceptualizing the Everyday Life of Native Americans in the Distant Past", *North American Archaeologist*, 25(4): 321-336.

Index

adoption 141, 160
aerial photographs 33, 34
Africa 4, 11, 12, 102, 122-123, 126, 128, 154
agriculture 72, 73, 75
agricultural landscapes 73
Alex Haley 1, 2
amateur genealogists 2, 3, 25, 108, 120, 137, 138, 179
American International Society of Genetic Genealogy 156
American Revolution 31, 34, 35
Americans
 African 3, 4, 15, 117, 121, 122, 129, 155, 158-159, 162-163
 Irish 13
 Native 6, 15, 155
 White 3, 14
ancestors *see* ancestry
ancestral homes 4, 116, 121, 175
 see also homeland
ancestral mythology 176
ancestral villages 3, 116, 129, 159, 167
 see also homeland
ancestry 1, 5, 11, 24, 37, 40, 43, 103, 104, 107, 137-148, 138, 142, 145, 147, 153, 154, 159, 160, 163, 166, 175, 176, 177, 180, 182
 disgraced 4
Anglo-Canadians 7
Antarctica 32
anthropologists 156, 183
anthropology 13
antiques 14, 120
 see also memorabilia
Arabs 165
archaeological evidence 165
archaeological sites 77, 115

archives 2, 7, 8, 14, 25, 45, 47, 48, 54, 67, 93, 99-109, 120, 124, 125, 162, 175, 181
aristocracy 3, 5, 10
Asia 102, 126
atlases *see* maps
Australia 23, 24, 117
autobiography 1, 15, 104-107, 148, 176

Balkans 9, 163
baptisms for the dead 3, 128-148
belonging, sense of 4, 5
Bible 9, 141, 142
 Genesis 9
biologists 156
bi-racism *see* racism
births 56, 57, 85, 104
birth certificates 102
birthright 1, 9
'blood quantum' 6
Blut und Boden (blood and soil) 13, 165
Book of Mormon 141-142
'book of remembrance' 148
boundaries/borders 25, 27, 29, 32, 34, 36, 37-39, 43, 45, 49, 52, 57, 99, 168, 183
Brazil 129, 154
'brick wall, the' 154, 179
Britain 4, 5, 24, 26, 34, 38, 84, 158, 164-165, 167
 see also England, Scotland, Wales

cadastral maps *see* maps
Canada 9, 23, 26, 36, 86, 127
 Alberta 180
 Quebec 7, 36, 117, 179
 Maritimes 36
 Ontario 36
Caribbean 102
cartography *see* maps
Catholicism 10, 12, 177

cemeteries 8, 11, 14, 28, 33, 35, 36, 120, 124, 125, 128
censorship 109
census 5, 23, 37, 100, 102, 107, 145, 167
childbearing 10
children 148, 176
China 127
Christianity 137-148
   see also Catholicism, Church of Jesus Christ of Latter-day Saints
churches 36, 37, 69, 103, 115, 120, 125, 160, 175, 182
church records see archives, parish records
Church of Jesus Christ of Latter-day Saints 3, 12, 32, 100, 106, 107, 119, 122, 137-148, 180
citizenship 1, 5, 6, 9, 164, 175
   see also nationality
Civil War see U.S. Civil War
clans 157, 166, 167
class see socio-economic class
coats-of-arms 5, 8, 167
collective memory 103-105, 108-109
colonial societies 2, 10, 14, 25, 45, 52, 156, 160
Confederate South 3
conferences 25, 33, 39
conservation 115
contested place 154
Croatia 127
cultural geographers 5, 160, 182
cultural geography see cultural geography
cultural landscapes 64, 65, 71, 115, 119, 128, 176
   see also historic landscapes
cultural resource management 65-66, 175
cultural traits 157
cyberspace 153, 160, 175
   democratization effects of 153
   see also Internet
cyber ethnic communities 175
Cyprus 121

Darwinism 4
Daughters of the American Revolution 11
death 10, 29, 40, 56, 57, 85, 86, 87, 94, 104
   see also funerals
death certificates 100, 102
'deep genealogy' see genetic genealogy

Denmark 13, 167
deterritorialization 161
diasporas 2, 116-118, 123, 137, 153, 161, 182-183
digital archives see archives
disabilities 87-95
diseases see illnesses
DNA testing see genetics
doctrine of genealogy 122

economists 183
Egypt 115
email 100, 108, 155, 180
   see also Internet
employment 29, 167
England 32, 115, 125, 158, 164
   see also Britain
environmental determinism 4
environmental impact assessment 63-78, 175
Estonians 164
ethnic group see ethnicity
ethnicity 1, 2, 4, 5, 7, 8, 9, 12, 13, 73, 116, 153, 158, 160, 161-164, 166
ethnocentricity 3, 153
   see also racism
Europe 8, 23, 24, 31, 38, 100, 101, 102, 115, 117, 119, 126, 157, 159, 165

family 1, 13, 23, 83-95, 100, 101, 103, 140, 158, 162, 180
   eternal nature of (Mormon doctrine) 137-148
family group sheets 28
family history, definition of 1
family history as school project/educational tool 3, 139
family history centers 32, 38, 100-101, 139, 147
   see also genealogical libraries
family history magazines 14
family trees 3, 5, 43, 107, 155
FamilySearch.org 145
Federal Transportation Act 64
Federation of Eastern European Family History Societies 38
females see women
Finland 39
food see traditional foods

France 7, 34, 35, 117, 163, 168
French Canadians 7, 10, 12
funerals 10

Gambia, The 2, 124
gate-keeping/gatekeepers 1, 6, 8, 83-95, 175
gazetteers 25, 26, 27, 29, 30, 32, 39, 49
gender 9, 101, 155, 158, 163, 166, 176
    see also women
genealogical libraries 14, 33, 119, 124, 125, 139, 144, 145, 146
    see also family history centers
genealogical societies 3, 178
Genealogical Society of Pennsylvania 145
Genealogical Society of Utah 32, 145-147
genealogy, definition of 1
genealogy, domain of the rich or white race 2
genealogy tourism see tourism
genetic distance 12
genetic genealogy 2, 153-169
genetic predispositions 9
geneticists 157, 158
genetics 10, 13, 15, 118, 153-169, 175
genome projects 153
geographic education 24
Geographic Information Systems 15, 35, 38, 43-57, 181
geography's ability to inform genetic genealogy 157-158
geomatics 1
Germany 6, 12, 38, 71, 125, 127, 179
Ghana 12, 122, 124
GIS see Geographic Information Systems
global positioning systems 28
globalization 108
Google Earth 30, 40
Google Maps 29
government agencies 120, 126-128, 175
GPS see global positioning systems
graves see cemeteries

haplogroups 156, 163, 164, 166, 168
health issues see illnesses
heritage 8, 64, 72
    industrial 68
heritage tourism see tourism
heterosexual norms 3
highways 64-65, 66, 77

historians 158, 159, 178, 183
historic landscapes 63
historical geographers see historical geography
historical geography 13, 44, 45, 48, 49, 51-52, 54, 57, 179
historical memory 67, 68, 71
historical societies 45, 55, 69-70, 104
Hitler 6
hobby see leisure, family history and genealogy as
homeland 2, 5, 7, 10, 11-13, 83, 89-90, 93, 116-129, 137, 158, 159, 160, 161, 168, 175
Homesteading Act of 1862 35
homophobia 3
    see also heterosexual norms
human genome projects 12
human geography 1
human origins 154
Hungary 127
hybridity 5, 154
    see also post-diasporic communities

Iberia 12, 163
Iceland 165
identity 1, 2, 4, 5, 6, 8, 10, 13, 78, 99, 103, 105, 116, 117, 119, 121, 129, 137, 158-162, 175, 180
    ethnic 9
    markers of 9
    place-based 12, 158-160
illegitimate children 6, 34, 166
illnesses 9, 10, 154
'imagined communities' 160-162, 166-167
immigrants see migration
immigration see migration
indigenous populations 158, 161, 163, 164
Industrial Revolution 37
inheritance 35
inheritance law 9
intermarriage 10, 161
    see also marriage
Internet 2, 7, 14, 26, 30, 36, 39, 45, 69, 99-109, 124, 153, 160, 162, 169, 175, 179, 182
Ireland 4, 11, 158, 163, 164, 179
Islam 138
Israel 9, 127-128

Italy 127

Jews 6, 9, 12, 122, 128, 165
Jim Crow laws 6
journal keeping 147-148

kinship 1, 3, 5, 7, 13, 88, 100, 101, 105, 123, 162, 166-167, 169, 182
kinship networks *see* kinship

land deeds *see* land records
land grants *see* land records
land ownership *see* property ownership
land records 15, 31, 34, 35, 49, 52-53, 55, 56, 128
landscape ecology 46
landscapes of memory 1, 160
language 12, 145, 168
  barriers 13
Lapps 164
Latvians 164
leisure
  family history and genealogy as 1, 2, 7, 137, 153, 179
letters 2, 83-95, 180, 182
libraries 27, 45, 47, 104
  see also genealogical libraries
Library of Congress 26, 34, 36, 39
life histories *see* autobiographies

Macedonians 164
Mali 2
maps 2, 11, 12, 15, 23-40, 45-49, 88, 166, 168, 178
  land and property 34-37, 45, 48, 49, 54, 125
  topographic 26, 27, 31, 43
marriage 6, 9, 28, 29, 30, 40, 57, 85, 86, 92, 94, 104, 141, 142, 167
  see also intermarriage
marriage certificates 100, 102, 120
mass culture 2
material culture 69
material possessions *see* memorabilia
Mayflower 11, 145
memorabilia 2, 6, 11, 14
  see also antiques
memory *see* identity and landscapes of memory
metes and bounds 34, 52

see also boundaries/borders
microfilms/microfiche 7, 52, 100-101, 145-146
migration 2, 5, 6, 7, 9, 10, 13, 27, 56, 63, 68, 70, 83-95, 107, 117, 129, 153, 154, 158, 166, 167, 175, 178, 179, 182
  routes 54
military conquest 8
mining 73, 83
minorities 2, 5, 13
  see also ethnicity, race
Mississippi River 143
mobility 15, 115-129, 167
Mormons *see* Church of Jesus Christ of Latter-day Saints
museums 15, 45, 47, 49, 115, 120, 121, 125
Muslims *see* Islam

National Environmental Policy Act 64, 66
National Geographic Society 157, 168
National Historic Preservation Act 64, 66
National Register of Historic Places 65, 77
National Society of Genealogy 25
nationality 4, 11-13, 153, 160, 162, 164, 168
  see also ethnicity, homeland
Netherlands 13, 32
New England Historical and Genealogical Society 145
'new family history' 5
'newbie' 155
newspapers 102, 107
New Zealand 118
New Testament 140
  See also Bible
New York Genealogical and Biographical Society 145
Nigeria 124
North America 23, 103, 117, 118, 119
  see also Canada, United States
Northern Ireland 9, 154
nostalgia 8, 64, 118, 119

'Old Genealogy' 13
Old Testament 142, 180
  See also Bible
online resources *see* Internet
'open identity' 5
oral history 2, 69, 103-104

Palestine 9
parish records 7, 12, 145
  see also archives
patriotism 161
patronymy 8
  see also surnames
pedigree charts 29
pedigrees 2, 7, 23, 25, 145, 159, 166
personal choice 9
personal heritage 3, 54, 101, 115-129, 137, 147-148, 175-183
Peru 115
photographs 2, 4, 15, 29, 44, 86, 88, 94, 102, 120, 175, 179, 181
pictures *see* photographs
pilgrim 122
pilgrimage 138-139
Pilgrims, the 2, 11
place-based heritage 9
place names 15, 30-34, 38, 48
podcasts 39
Poland 127
political incorrectness 8
political scientists 183
political use of genealogy 13
postcolonial *see* colonial societies
post-diasporic communities 8
  see also hybridity
post-ethnic space 14
power relations 8, 9
prejudices 10, 160
  see also racism
primordialists 157, 165
  see also 'Blut und Boden'
professional genealogists 3, 4, 24, 120
property ownership 37, 43
psychologists 183

race 6, 13, 116
racism 6, 14, 129, 154
  see also prejudices
relatives 148
  see also ancestry
religion 9, 10, 13, 14, 28, 34, 116, 122, 137-148, 175, 176
religious affiliation 9, 160, 179
religious freedom 29
religious ordinances/rites 140-148, 175
residential location 2

reunions 2, 148
rootlessness 11, 116-118, 120
  see also identity
roots *see* ancestry, heritage
'roots tourism' *see* tourism

Saami *see* Lapps
satellite imagery 30
Scandinavia 8, 39
schools 24, 37, 161
Scotland 11, 14, 83-95, 118, 123, 127
Scottish tartans 11, 167
sectarian conflicts 4
segregation 14
  see also racism, prejudices
septs *see* clans
'serious genealogists' *see* professional genealogists, amateur genealogists
sexism 3, 8
  see also heterosexual norms
sexuality 10
slavery 121-124, 128
social boundaries
  construction of 5
socio-economic class 1, 5, 9, 11, 13
sociologists 183
sociology 13
Soviet Union 38
Spain 34, 35, 168
spatial correlation 153, 166
spirituality 175
surname projects 158, 159, 160, 166, 168, 169, 179
surnames 7, 8
  see also patronymy
Switzerland 127, 182

tartans *see* Scottish tartans
technology 99-109
temples 138-148, 180
territory 162, 163
  see also homeland
Thomas Jefferson 34, 163
toponyms *see* place names
topophilia 120-121
tour operators 14, 120, 123-126
tourism 1, 2, 11, 12, 14, 15, 63, 115-129, 138, 159, 161, 175, 180
tourists *see* tourism

traditional foods 15, 161, 168
transnationalism 162
  see also diaspora, identity
transportation 47, 49, 123
transportation networks 35
transportation routes 43
travel *see* tourism

United Kingdom *see* Britain
United Nations 103
United States of America 2, 3, 6, 11, 24, 27, 31, 35, 36, 44-45, 102, 105, 127, 154, 161, 179, 182
  Alabama 26
  Arizona 35, 182
  Arkansas 177
  California 35
  Florida 35, 129
  Hawaii 34
  Idaho 180
  Illinois 23, 143
  Indiana 23, 34
  Kentucky 34
  Louisiana 35, 48
  Maine 34
  Maryland 23, 46, 74
  Massachusetts 11
  Mississippi 177
  Missouri 23
  New England 10, 24
  New Mexico 35
  New York 37, 140, 143
  North Carolina 44-57, 181
  Ohio 23, 143, 176-177
  Pennsylvania 63, 83
  Tennessee 34
  Texas 34
  Utah 32, 38, 119, 139-140, 143-144, 182
  Vermont 34
  Virginia 23, 46, 163
  Washington, DC 26, 35, 180
  West Virginia 23, 34
US Census Bureau 23
US Civil War 11, 37
US Constitution 8
US Federal Highway Administration 63
US Geological Survey 32, 43

Venezuela 129
victimization 9, 13
Viking genes 158, 165, 166
virtual spaces 175

Wales 8, 127, 182
websites 2, 12, 28, 32, 36, 39, 107, 126, 155-156, 160
  see also Internet
weddings 10
  see also marriage
women 4, 9, 28, 83-95, 156
  see also gender, male
World Bank 103
World War I 24, 36
World War II 119

xenophobia 163
  see also prejudice, racism, ethnocentricity